# Mathematics for Everyday Life 11

Enzo Carli
Sandra Emms Jones
Alexis Galvao
Loraine Wilson

Toronto, Canada

National Library of Canada Cataloguing in Publication

Main entry under title:
Mathematics for everyday life 11/Enzo Gildo Carli ... [et al.].

Includes index.
ISBN 0-7725-2924-8

1. Finance, Personal—Mathematics. I. Carli, E. G

QA39.2.M293 2002 332.024'001'513 C2002-901879-X



We acknowledge for their financial support of our publishing program, the Canada Council, the Ontario Arts Council, and the Government of Canada through the Book Publishing Industry Development Program (BPIDP).

Publisher: Jan Elliott
Managing Editor: Janice Nixon
Developmental Editors: Janice Nixon, Mary Reeve
Copy/Production Editor: Kate Revington
Editorial Assistant: Andrea Byrne
Cover and Text Design: Dave Murphy/ArtPlus Ltd.
Page Layout: Miriam Semple, Matthew Coyle/ArtPlus Ltd.
Technical Art: ArtPlus Ltd.
Photographer: Trent Photographics
Photo Research: Lisa Brant

For more information contact
Nelson
1120 Birchmount Road,
Scarborough, Ontario, M1K 5G4.
Or you can visit our Internet site at
http://www.nelson.com

Printed and bound in Canada
5 6 7 11 10 09

## Acknowledgements

The authors and editors of *Mathematics for Everyday Life 11* wish to thank the reviewers listed below for their valuable input and assistance in ensuring that this text meets the needs of teachers and students in Ontario.

Tom Chapman
Hastings and Prince Edward District School Board

Chris Dearling
Burlington

Peter Joong
Toronto District School Board

Linda Palmason
Kawartha Pine Ridge District School Board

Peter Saarimaki
Scarborough

Shirley Scott
District School Board of Niagara

Susan K. Smith
Peel District School Board

# Contents

## Chapter 6 *Banking Transactions and Saving Money*



## Chapter 7 *Investing Money*



## Chapter 8 *Taking a Trip*



## Chapter 9 *Borrowing Money*



## Chapter 10 *Buying a Car*

# About Your Textbook

When you turn the pages of your new textbook, you will notice that most sections begin with **Explore**. Here, you usually work in a small group or with a partner. You are given an opportunity to connect your prior math knowledge and your personal experience to concepts in the section. What you have to do varies in scope from briefly reflecting and discussing to applying the steps of an inquiry/problem solving process.

An **inquiry/problem solving process** involves
- formulating questions
- selecting strategies, resources, technology, and tools
- representing in mathematical form
- interpreting information and forming conclusions
- reflecting on the reasonableness of results

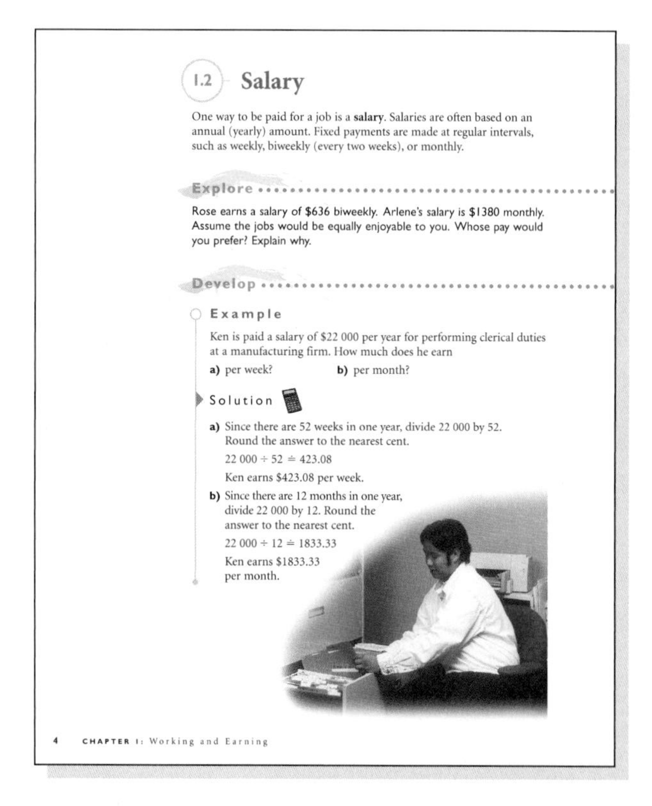
1.2 Salary

One way to be paid for a job is a **salary**. Salaries are often based on an annual (yearly) amount. Fixed payments are made at regular intervals, such as weekly, biweekly (every two weeks), or monthly.

Explore

Rose earns a salary of $636 biweekly. Arlene's salary is $1380 monthly. Assume the jobs would be equally enjoyable to you. Whose pay would you prefer? Explain why.

Develop

Example

Ken is paid a salary of $22 000 per year for performing clerical duties at a manufacturing firm. How much does he earn

**a)** per week? **b)** per month?

Solution

**a)** Since there are 52 weeks in one year, divide 22 000 by 52. Round the answer to the nearest cent.

22 000 ÷ 52 ≐ 423.08

Ken earns $423.08 per week.

**b)** Since there are 12 months in one year, divide 22 000 by 12. Round the answer to the nearest cent.

22 000 ÷ 12 ≐ 1833.33

Ken earns $1833.33 per month.

4 CHAPTER 1: Working and Earning

Most sections continue with **Develop**.

Here, you sometimes work in a small group or with a partner. The concepts of the section are developed by presenting you with
- directed questions to answer and/or
- examples and fully worked solutions to follow

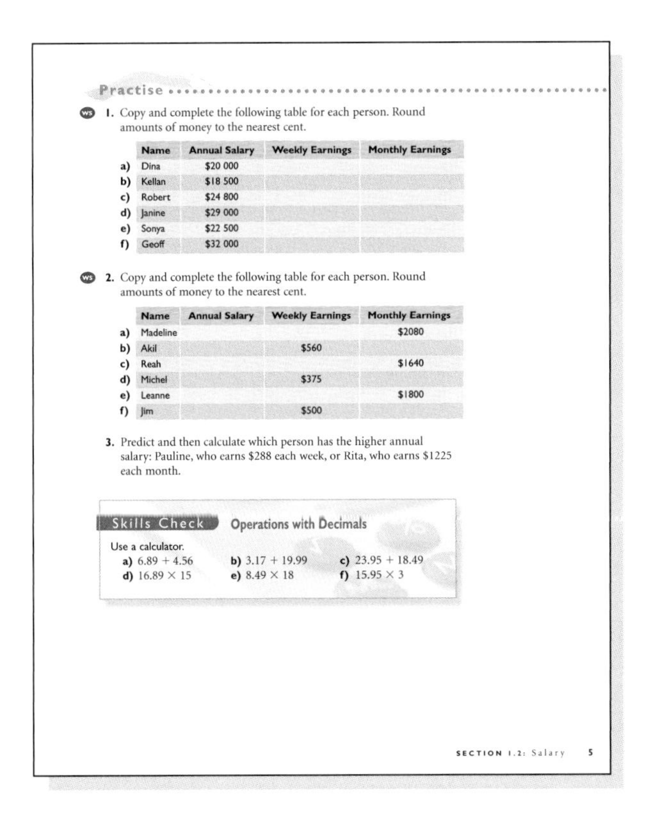

Practise

1. Copy and complete the following table for each person. Round amounts of money to the nearest cent.

| | Name | Annual Salary | Weekly Earnings | Monthly Earnings |
|---|---|---|---|---|
| a) | Dina | $20 000 | | |
| b) | Kellan | $18 500 | | |
| c) | Robert | $24 800 | | |
| d) | Janine | $29 000 | | |
| e) | Sonya | $22 500 | | |
| f) | Geoff | $32 000 | | |

2. Copy and complete the following table for each person. Round amounts of money to the nearest cent.

| | Name | Annual Salary | Weekly Earnings | Monthly Earnings |
|---|---|---|---|---|
| a) | Madeline | | | $2080 |
| b) | Akil | | $560 | |
| c) | Reah | | | $1640 |
| d) | Michel | | $375 | |
| e) | Leanne | | | $1800 |
| f) | Jim | | $500 | |

3. Predict and then calculate which person has the higher annual salary: Pauline, who earns $288 each week, or Rita, who earns $1225 each month.

Skills Check — Operations with Decimals

Use a calculator.

a) 6.89 + 4.56 b) 3.17 + 19.99 c) 23.95 + 18.49
d) 16.89 × 15 e) 8.49 × 18 f) 15.95 × 3

SECTION 1.2: Salary 5

Most sections then have **Practise**. Here, you frequently work alone to answer questions about the skills and concepts presented in the section.

Some sections include a **Skills Check**. Here, you review skills that you need for success in upcoming sections.

# In each chapter you will find one of each of the following special sections.

*Chapter Opener*

This introduction outlines what you are expected to accomplish in the chapter.

CHAPTER 6 Banking Transactions and Saving Money

In this chapter you will
- identify banking transactions
- interpret and check bank statements and passbooks
- investigate savings alternatives available at banks
- calculate simple interest and amount
- calculate amount and compound interest
- determine the effects of compound interest on saving a small sum on a regular basis
- identify the effects of different compounding periods

Near the end of the chapter, for the earnings, expenses, and savings related to "your job," you will
- determine banking transactions that apply
- estimate the monthly service charge that will apply
- choose an appropriate savings alternative
- determine the amount of a Guaranteed Investment Certificate you have received

4.5 Career Focus: Sales and Merchandising Clerk

Sandy is a clerk in a hardware store. One task that she enjoys is creating signs for end-of-aisle displays. The policy of the store is to have a sign for sale items on end-of-aisle displays featuring this information:
- the name of the item
- the sale price
- the regular price
- the discount rate
- the stock item number

The item name and the sale price always appear in the largest print.

Sandy created this sign using computer software.



1. Create a sign that meets the specifications given above for each sale. Use a word processor or other computer software, or draw by hand.
   a) bicycle helmets
      regular price $24.99
      discount rate 15% off
      small #10-365 medium #10-366 large #10-367
   b) watering cans
      regular price $2.49
      discount rate half price
      #33-3487
   c) garden shovels
      regular price $9.79
      discount rate 10% off
      #33-4987
2. On your sign, what is the ratio of the height of the largest letters to the height of the smallest letters?

SECTION 4.5: Career Focus: Sales and Merchandising Clerk 73

3. Another job that Sandy enjoys is pricing new merchandise. The selling price is the cost to the store plus a markup. The **markup** rate is a percent of the cost to the store. Stores mark up their costs to pay their overhead, such as building maintenance, rent or taxes, utilities, salaries and benefits for employees and, of course, to make a profit.

   The following is a list of new merchandise at the hardware store. The markup code is explained below the list.

| Item | Stock Item # | Cost per Item | Markup Code |
|---|---|---|---|
| hammer | 73-2318 | $9.24 | R |
| sandpaper | 73-8719 | $0.19 | R |
| measuring tape | 73-3240 | $2.12 | S |
| door stop | 20-7493 | $0.78 | R |
| window cleaner | 42-3938 | $1.23 | T |
| picture wire | 22-4829 | $0.84 | R |
| pack of sponges | 42-0820 | $0.47 | T |
| air freshener | 42-2767 | $1.02 | T |
| plant food | 33-5938 | $2.24 | T |
| broom | 42-5720 | $1.35 | S |
| package of tacks | 22-5934 | $0.28 | S |
| Markup Code (% of cost): R 35% S 50% T 65% | | | |

   Sandy determines the selling price of the new model of hammers by one of two methods.

   **Method 1**
   Find 35% of $9.24, add the amount to $9.24, and round up to the next 9¢.

   **Method 2**
   Find 135% of $9.24 and round up to the next 9¢.

   a) Find the selling price of the new hammer using both methods. Which do you prefer? Why?
   b) Why do you think the selling price is rounded up to the next 9¢?
   c) Use the method of your choice. Find the selling prices for the other items of new merchandise listed above.
4. Sandy also works on the cash register. She makes change for customers' purchases. Determine the bills and coins she should give to the customer of each purchase for the amounts tendered.
   a) total price of $13.78, when $15 tendered
   b) total price of $20.67, when $22.17 tendered
   c) total price of $26.28, when $31.30 tendered
   d) total price of $32.52, when $40 tendered
   e) total price of $8.93, when $10.03 tendered

74 CHAPTER 4: Making Purchases

*Career Focus*

In this section you read and answer questions about a job which can be entered right after high school. The questions that you answer also offer practice in some of the Essential Skills—

- text reading
- document use
- writing
- numeracy (math)
- working with others
- continuous learning
- computer skills
- thinking skills

Developing these skills will enable you to participate fully in the workplace and the community.

*Putting It All Together*

In this section you are given an opportunity to demonstrate skills in the interconnected concepts of the chapter. Sometimes, you are presented with specific questions to answer. Other times, you are presented with a problem that calls upon you to apply the steps of an inquiry/problem solving process.

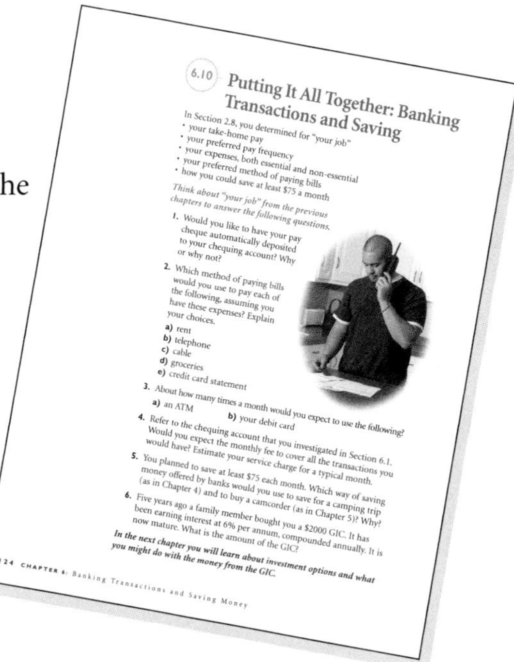

6.10 Putting It All Together: Banking Transactions and Saving

In Section 2.8, you determined for "your job"
- your take-home pay
- your preferred pay frequency
- your expenses, both essential and non-essential
- your preferred method of paying bills
- how you could save at least $75 a month

*Think about "your job" from the previous chapters to answer the following questions.*

1. Would you like to have your pay cheque automatically deposited to your chequing account? Why or why not?
2. Which method of paying bills would you use to pay each of the following, assuming you have these expenses? Explain your choices.
   a) rent
   b) telephone
   c) cable
   d) groceries
   e) credit card statement
3. About how many times a month would you expect to use the following?
   a) an ATM  b) your debit card
4. Refer to the chequing account that you investigated in Section 6.1. Would you expect the monthly fee to cover all the transactions you would have? Estimate your service charge for a typical month.
5. You planned to save at least $75 each month. Which way of saving money offered by banks would you use to save for a camping trip (as in Chapter 4) and to buy a camcorder (as in Chapter 5)? Why?
6. Five years ago a family member bought you a $2000 GIC. It has been earning interest at 6% per annum, compounded annually. It is now mature. What is the amount of the GIC?

***In the next chapter you will learn about investment options and what you might do with the money from the GIC.***

124 CHAPTER 6: Banking Transactions and Saving Money

*Chapter Review*

Here, you find questions similar to the Practise questions in earlier sections. Answering them helps you prepare for a chapter test.

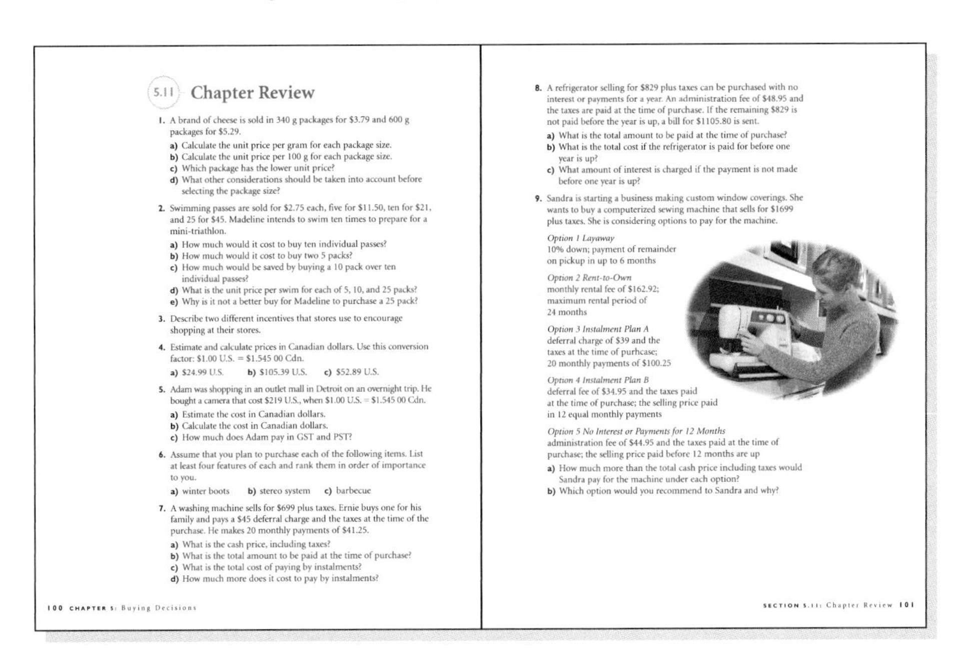

5.11 Chapter Review

1. A brand of cheese is sold in 340 g packages for $3.79 and 600 g packages for $5.29.
   a) Calculate the unit price per gram for each package size.
   b) Calculate the unit price per 100 g for each package size.
   c) Which package has the lower unit price?
   d) What other considerations should be taken into account before selecting the package size?
2. Swimming passes are sold for $2.75 each, five for $11.50, ten for $21, and 25 for $45. Madeline intends to swim ten times to prepare for a mini-triathlon.
   a) How much would it cost to buy ten individual passes?
   b) How much would it cost to buy two 5 packs?
   c) How much would be saved by buying a 10 pack over ten individual passes?
   d) What is the unit price per swim for each of 5, 10, and 25 packs?
   e) Why is it not a better buy for Madeline to purchase a 25 pack?
3. Describe two different incentives that stores use to encourage shopping at their stores.
4. Estimate and calculate prices in Canadian dollars. Use this conversion factor: $1.00 U.S. = $1.545 00 Cdn.
   a) $24.99 U.S.  b) $105.39 U.S.  c) $52.89 U.S.
5. Adam was shopping in an outlet mall in Detroit on an overnight trip. He bought a camera that cost $219 U.S., when $1.00 U.S. = $1.545 00 Cdn.
   a) Estimate the cost in Canadian dollars.
   b) Calculate the cost in Canadian dollars.
   c) How much does Adam pay in GST and PST?
6. Assume that you plan to purchase each of the following items. List at least four features of each and rank them in order of importance to you.
   a) winter boots  b) stereo system  c) barbecue
7. A washing machine sells for $699 plus taxes. Ernie buys one for his family and pays a $45 deferral charge and the taxes at the time of the purchase. He makes 20 monthly payments of $41.25.
   a) What is the cash price, including taxes?
   b) What is the total amount to be paid at the time of purchase?
   c) What is the total cost of paying by instalments?
   d) How much more does it cost to pay by instalments?

100 CHAPTER 5: Buying Decisions

8. A refrigerator selling for $829 plus taxes can be purchased with no interest or payments for a year. An administration fee of $48.95 and the taxes are paid at the time of purchase. If the remaining $829 is not paid before the year is up, a bill for $1105.80 is sent.
   a) What is the total amount to be paid at the time of purchase?
   b) What is the total cost if the refrigerator is paid for before one year is up?
   c) What amount of interest is charged if the payment is not made before one year is up?
9. Sandra is starting a business making custom window coverings. She wants to buy a computerized sewing machine that sells for $1699 plus taxes. She is considering options to pay for the machine.

   *Option 1 Layaway*
   10% down; payment of remainder on pickup in up to 6 months

   *Option 2 Rent-to-Own*
   monthly rental fee of $162.92; maximum rental period of 24 months

   *Option 3 Instalment Plan A*
   deferral charge of $39 and the taxes at the time of purhcase; 20 monthly payments of $100.25

   *Option 4 Instalment Plan B*
   deferral fee of $34.95 and the taxes paid at the time of purchase; the selling price paid in 12 equal monthly payments

   *Option 5 No Interest or Payments for 12 Months*
   administration fee of $44.95 and the taxes paid at the time of purchase; the selling price paid before 12 months are up

   a) How much more than the total cash price including taxes would Sandra pay for the machine under each option?
   b) Which option would you recommend to Sandra and why?

SECTION 5.11: Chapter Review 101

# Technology in *Mathematics for Everyday Life 11*

Technology plays an important role in supporting your learning.

You will use

- calculators to perform operations with decimals
- spreadsheets to perform repetitive calculations and draw graphs

Using these tools will enable you to focus on the concepts that you are learning about.

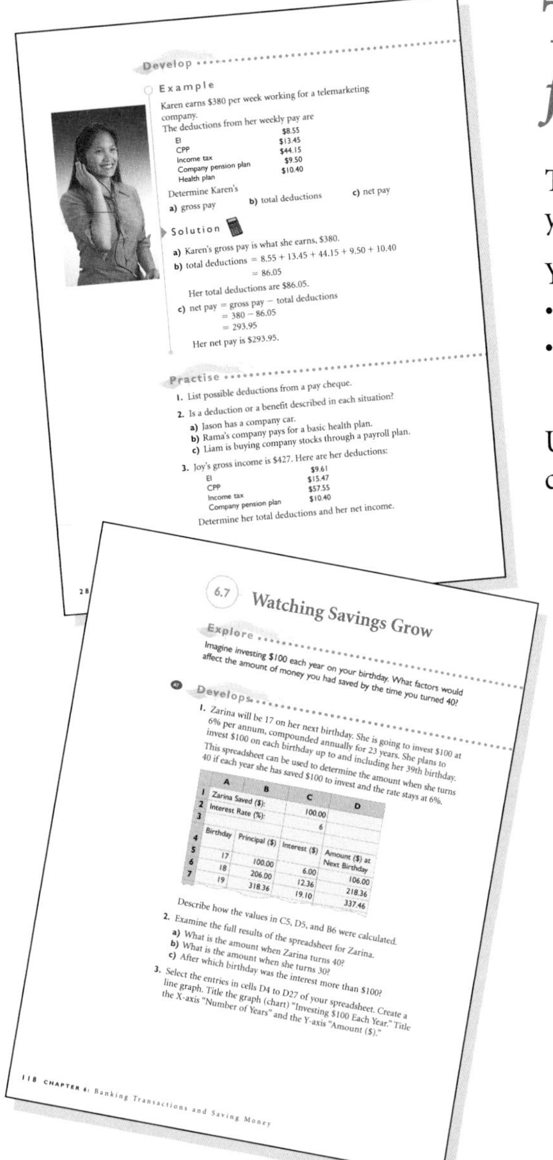

**Develop**

**Example**

Karen earns $380 per week working for a telemarketing company.
The deductions from her weekly pay are

| | |
|---|---|
| EI | $8.55 |
| CPP | $13.45 |
| Income tax | $44.15 |
| Company pension plan | $9.50 |
| Health plan | $10.40 |

Determine Karen's

**a)** gross pay **b)** total deductions **c)** net pay

**Solution**

**a)** Karen's gross pay is what she earns, $380.

**b)** total deductions = 8.55 + 13.45 + 44.15 + 9.50 + 10.40
= 86.05

Her total deductions are $86.05.

**c)** net pay = gross pay − total deductions
= 380 − 86.05
= 293.95

Her net pay is $293.95.

**Practise**

**1.** List possible deductions from a pay cheque.

**2.** Is a deduction or a benefit described in each situation?

**a)** Jason has a company car.
**b)** Rama's company pays for a basic health plan.
**c)** Liam is buying company stocks through a payroll plan.

**3.** Joy's gross income is $427. Here are her deductions:

| | |
|---|---|
| EI | $9.61 |
| CPP | $15.47 |
| Income tax | $57.55 |
| Company pension plan | $10.40 |

Determine her total deductions and her net income.

28

## 6.7 Watching Savings Grow

**Explore**

Imagine investing $100 each year on your birthday. What factors would affect the amount of money you had saved by the time you turned 40?

**Develop**

**1.** Zarina will be 17 on her next birthday. She is going to invest $100 at 6% per annum, compounded annually for 23 years. She plans to invest $100 on each birthday up to and including her 39th birthday. This spreadsheet can be used to determine the amount when she turns 40 if each year she has saved $100 to invest and the rate stays at 6%.

| | A | B | C | D |
|---|---|---|---|---|
| 1 | Zarina Saved ($): | | 100.00 | |
| 2 | Interest Rate (%): | | 6 | |
| 3 | | | | |
| 4 | Birthday | Principal ($) | Interest ($) | Amount ($) at Next Birthday |
| 5 | 17 | 100.00 | 6.00 | 106.00 |
| 6 | 18 | 206.00 | 12.36 | 218.36 |
| 7 | 19 | 318.36 | 19.10 | 337.46 |

Describe how the values in C5, D5, and B6 were calculated.

**2.** Examine the full results of the spreadsheet for Zarina.

**a)** What is the amount when Zarina turns 40?
**b)** What is the amount when she turns 30?
**c)** After which birthday was the interest more than $100?

**3.** Select the entries in cells D4 to D27 of your spreadsheet. Create a line graph. Title the graph (chart) "Investing $100 Each Year." Title the X-axis "Number of Years" and the Y-axis "Amount ($)."

118 CHAPTER 6: Banking Transactions and Saving Money

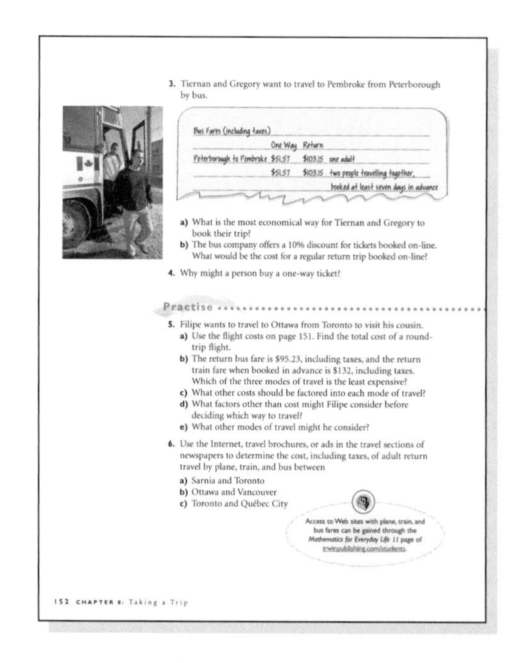

**3.** Tiernan and Gregory want to travel to Pembroke from Peterborough by bus.

Bus Fares (including taxes)

| | One Way | Return | |
|---|---|---|---|
| Peterborough to Pembroke | $51.57 | $103.15 | one adult |
| | $51.57 | $103.15 | two people travelling together, booked at least seven days in advance |

**a)** What is the most economical way for Tiernan and Gregory to book their trip?
**b)** The bus company offers a 10% discount for tickets booked on-line. What would be the cost for a regular return trip booked on-line?

**4.** Why might a person buy a one-way ticket?

**Practise**

**5.** Filipe wants to travel to Ottawa from Toronto to visit his cousin.

**a)** Use the flight costs on page 151. Find the total cost of a round-trip flight.
**b)** The return bus fare is $95.23, including taxes, and the return train fare when booked in advance is $132, including taxes. Which of the three modes of travel is the least expensive?
**c)** What other costs should be factored into each mode of travel?
**d)** What factors other than cost might Filipe consider before deciding which way to travel?
**e)** What other modes of travel might he consider?

**6.** Use the Internet, travel brochures, or ads in the travel sections of newspapers to determine the cost, including taxes, of adult return travel by plane, train, and bus between

**a)** Sarnia and Toronto
**b)** Ottawa and Vancouver
**c)** Toronto and Québec City

Access to Web sites with plane, train, and bus fares can be gained through the *Mathematics for Everyday Life 11* page of irwinpublishing.com/students.

152 CHAPTER 8: Taking a Trip

You will also use the Internet, as well as print sources, such as newspapers, flyers, pamphlets, and brochures about a variety of topics, to research current information.

# Watch for the following.

**AT** This indicates that an **Alternative to Technology** master is available. Your teacher may provide it if your class is unable to work on computers with spreadsheet software.

This tells you that you can get linked to relevant Web sites at the **Irwin Publishing Web site**.

**WS** This means that an organizational **Work Sheet** master is available. Your teacher may provide it to help you organize your work for a particular question or group of questions.

**A** This tells you that an **Assessment** master with a rubric is available. Your teacher may provide it to help you understand how your work might be assessed.

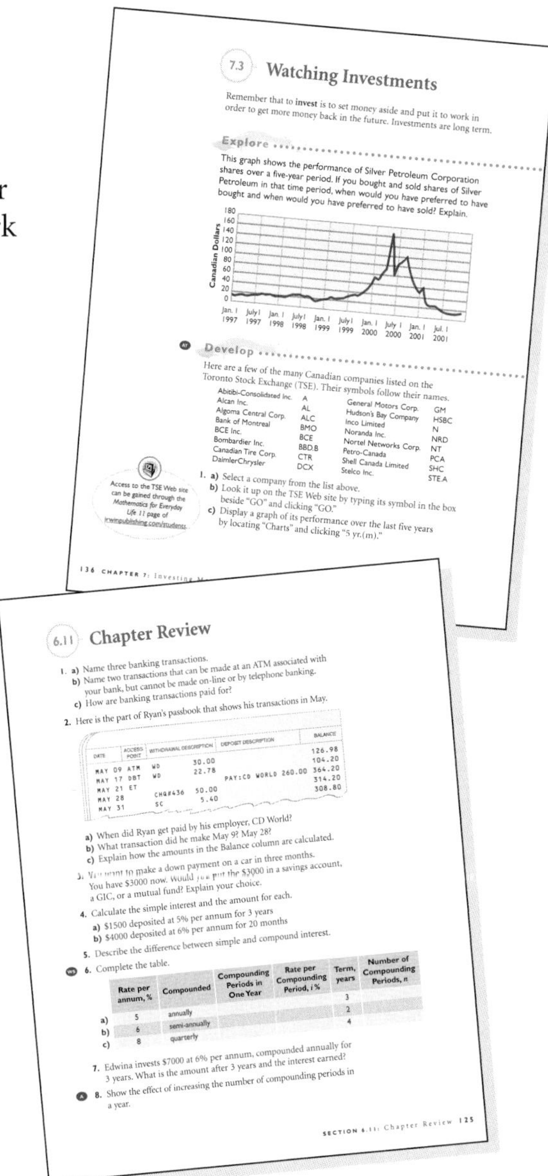

7.3 **Watching Investments**

Remember that to **invest** is to set money aside and put it to work in order to get more money back in the future. Investments are long term.

**Explore**

This graph shows the performance of Silver Petroleum Corporation shares over a five-year period. If you bought and sold shares of Silver Petroleum in that time period, when would you have preferred to have bought and when would you have preferred to have sold? Explain.



**AT** **Develop**

Here are a few of the many Canadian companies listed on the Toronto Stock Exchange (TSE). Their symbols follow their names.

| Company | Symbol | Company | Symbol |
|---|---|---|---|
| Abitibi-Consolidated Inc. | A | General Motors Corp. | GM |
| Alcan Inc. | AL | Hudson's Bay Company | HSBC |
| Algoma Central Corp. | ALC | Inco Limited | N |
| Bank of Montreal | BMO | Noranda Inc. | NRD |
| BCE Inc. | BCE | Nortel Networks Corp. | NT |
| Bombardier Inc. | BBD.B | Petro-Canada | PCA |
| Canadian Tire Corp. | CTR | Shell Canada Limited | SHC |
| DaimlerChrysler | DCX | Stelco Inc. | STE.A |

1. **a)** Select a company from the list above.
**b)** Look it up on the TSE Web site by typing its symbol in the box beside "GO" and clicking "GO."
**c)** Display a graph of its performance over the last five years by locating "Charts" and clicking "5 yr.(m)."

Access to the TSE Web site can be gained through the *Mathematics for Everyday Life 11* page of irwinpublishing.com/students.

136 CHAPTER 7: Investing M

6.11 **Chapter Review**

1. **a)** Name three banking transactions.
**b)** Name two transactions that can be made at an ATM associated with your bank, but cannot be made on-line or by telephone banking.
**c)** How are banking transactions paid for?

2. Here is the part of Ryan's passbook that shows his transactions in May.

| DATE | ACCESS POINT | WITHDRAWAL DESCRIPTION | | DEPOSIT DESCRIPTION | BALANCE |
|---|---|---|---|---|---|
| | | | | | 126.98 |
| MAY 09 | ATM | WD | 30.00 | | 104.20 |
| MAY 17 | DBT | WD | 22.78 | PAY:CD WORLD 260.00 | 364.20 |
| MAY 21 | ET | | | | 314.20 |
| MAY 28 | | CHQ#436 | 50.00 | | 308.80 |
| MAY 31 | | SC | 5.40 | | |

**a)** When did Ryan get paid by his employer, CD World?
**b)** What transaction did he make May 9? May 28?
**c)** Explain how the amounts in the Balance column are calculated.

3. You want to make a down payment on a car in three months. You have $3000 now. Would you put the $3000 in a savings account, a GIC, or a mutual fund? Explain your choice.

4. Calculate the simple interest and the amount for each.
**a)** $1500 deposited at 5% per annum for 3 years
**b)** $4000 deposited at 6% per annum for 20 months

5. Describe the difference between simple and compound interest.

**WS** 6. Complete the table.

| | Rate per annum, % | Compounded | Compounding Periods in One Year | Rate per Compounding Period, i % | Term, years | Number of Compounding Periods, n |
|---|---|---|---|---|---|---|
| a) | 5 | annually | | | 3 | |
| b) | 6 | semi-annually | | | 2 | |
| c) | 8 | quarterly | | | 4 | |

7. Edwina invests $7000 at 6% per annum, compounded annually for 3 years. What is the amount after 3 years and the interest earned?

**A** 8. Show the effect of increasing the number of compounding periods in a year.

SECTION 6.11: Chapter Review 125

# Other Features

## *CD with spreadsheet templates*

Templates in Microsoft Excel and Corel Quattro Pro are provided for you every time you need to work with spreadsheets.

## *Glossary*

All the new terms in the book are listed alphabetically. Terms are explained when first introduced, but whenever you need to confirm the meaning of a term, you can turn to the Glossary.

## *Answers*

You will find that answers to all questions are given except when they would vary depending on personal choices. Check your answers against these. If your answer differs, work backward to try to understand how the textbook's answer was determined.

CHAPTER

# 1 *Working and Earning*

In this chapter, you will solve problems involving the various ways people are paid for their work. These ways include

- salary
- piecework
- hourly rate and overtime rate
- hourly rate and tips
- commission

Near the end of the chapter, you will

- select a suitable job that you might hold now or in the near future
- apply for the job
- calculate how much you will be paid for the job weekly, monthly, and yearly

# 1.1 Finding a Job

## Explore

Chantel is in Grade 11. She hopes to find a full-time job after Grade 12. What could she be doing now while still attending school to help make finding a full-time job easier?

## Develop

WS **1. a)** Read the job ads on the next page. Discuss and record what you think the terms in red mean.

**b)** Compare your answers in part a) with the meanings given in the Glossary, which starts on page 206.

**2.** Some positions ask for a resume to be faxed or e-mailed. If you do not have the capability to fax or e-mail from your home, how can you apply for such positions?

WS **3.** For each of the following jobs from the ads,
- child-care worker
- sales help
- entry-level clerk
- server

answer these questions.

**a)** What personal characteristics are needed?
**b)** What education is needed?
**c)** What experience is needed?
**d)** How will the successful candidate be paid?
**e)** Are you qualified to apply?

## Child-Care Worker

Caring child-care worker required immediately for child-care facility in the north end of the city.
ECE certificate required.
Salary: $29 500/year
Excellent benefits package.
Fax resume to 555-4500.

## SUMMER CAMP COUNSELLORS REQUIRED FOR FOUR WEEKS AT CAMP IN SOUTHERN ITALY

Work with Canadian and U.S. students attending school at army bases in Europe. Fluency in Italian an asset, but not required. Room and board, travel allowance, health insurance, and $1000 U.S. stipend included.
E-mail resume to
yourname@irwinpublishing.com.

## *Sales Help Needed*

Friendly, conscientious workers required for the Retail River store in White Plains Mall.
Fluency in English is required. Fluency in a second language is an asset.
Earn $54 per shift plus commission.
No experience necessary. Will train.
Job application forms available at Retail River in White Plains Mall, Townsville.

## CLERICAL OPENINGS

Entry-level positions available at fast-growing computer graphics company.
Good communication and organizational skills required.
Must have experience with Excel and Word.
Salary: $24 000
Quarterly Bonus: $1200
Fax resume to 555-3490.

## Wait Staff Needed

Servers needed for a roadhouse opening close to the Community Centre.
Neat appearance and friendly personality a must.
Up to 20 hours weekly.
Minimum wage plus tips.
Pick up job application at site.

## Attention: Electricians

Electricians needed for new development south of the city.
Licensed union members only.
At least 3 years experience required.
Hourly rate depends on experience.
Possible overtime hours.
Fax resume to 555-8176.

## *Wooden Pallet Builders*

required at Sure-Wood Manufacturing Company.
Successful candidates will be paid by piecework —
35¢ to 50¢ per pallet with about 300 units being built per day.
Apply in person only: 9 a.m.–3 p.m.
86 Industrial Park Drive, Townsville.

## Practise

4. Use newspapers or the Internet to find two job ads that interest you.

Access to Web sites for jobs can be gained through the *Mathematics for Everyday Life 11* page of irwinpublishing.com/students.

WS 5. For each job that you found in question 4, answer these questions.
   a) What are the duties?
   b) Where is it located?
   c) How does it pay (e.g., salary, hourly rate, commission, piecework)?
   d) Is it part time or full time?
   e) What personal characteristics are needed?
   f) What education is needed?
   g) What experience is needed?
   h) To apply, who should be contacted?
   i) How should the person be contacted?

6. List three items of important information that your ads did not supply. Do not use items covered in question 5.

# 1.2 Salary

One way to be paid for a job is a **salary**. Salaries are often based on an annual (yearly) amount. Fixed payments are made at regular intervals, such as weekly, biweekly (every two weeks), or monthly.

## Explore

Rose earns a salary of $636 biweekly. Arlene's salary is $1380 monthly. Assume the jobs would be equally enjoyable to you. Whose pay would you prefer? Explain why.

## Develop

### Example

Ken is paid a salary of $22 000 per year for performing clerical duties at a manufacturing firm. How much does he earn

**a)** per week? **b)** per month?

### Solution

**a)** Since there are 52 weeks in one year, divide 22 000 by 52. Round the answer to the nearest cent.

$22\,000 \div 52 \doteq 423.08$

Ken earns $423.08 per week.

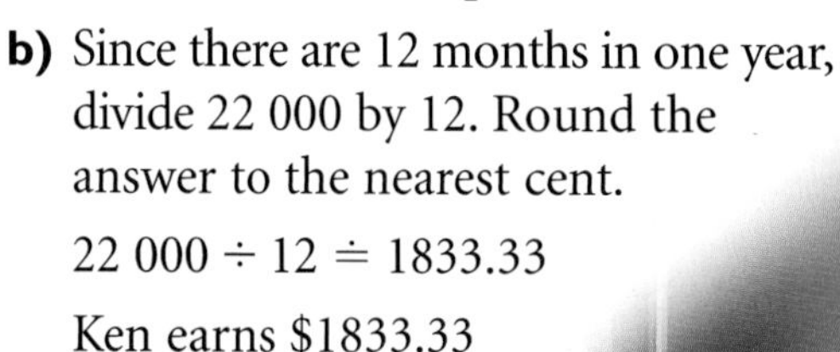

**b)** Since there are 12 months in one year, divide 22 000 by 12. Round the answer to the nearest cent.

$22\,000 \div 12 \doteq 1833.33$

Ken earns $1833.33 per month.

## Practise

WS **1.** Copy and complete the following table for each person. Round amounts of money to the nearest cent.

| | Name | Annual Salary | Weekly Earnings | Monthly Earnings |
|---|---|---|---|---|
| **a)** | Dina | $20 000 | | |
| **b)** | Kellan | $18 500 | | |
| **c)** | Robert | $24 800 | | |
| **d)** | Janine | $29 000 | | |
| **e)** | Sonya | $22 500 | | |
| **f)** | Geoff | $32 000 | | |

WS **2.** Copy and complete the following table for each person. Round amounts of money to the nearest cent.

| | Name | Annual Salary | Weekly Earnings | Monthly Earnings |
|---|---|---|---|---|
| **a)** | Madeline | | | $2080 |
| **b)** | Akil | | $560 | |
| **c)** | Reah | | | $1640 |
| **d)** | Michel | | $375 | |
| **e)** | Leanne | | | $1800 |
| **f)** | Jim | | $500 | |

**3.** Predict and then calculate which person has the higher annual salary: Pauline, who earns $288 each week, or Rita, who earns $1225 each month.

### Skills Check — Operations with Decimals

Use a calculator.

**a)** $6.89 + 4.56$ **b)** $3.17 + 19.99$ **c)** $23.95 + 18.49$

**d)** $16.89 \times 15$ **e)** $8.49 \times 18$ **f)** $15.95 \times 3$

# 1.3 Piecework

Some workers are paid by **piecework**, that is, they are paid based on each unit they produce or sell. Earnings may be based on **salary plus piecework** or **straight piecework**.

## Explore

You have been offered a part-time job assembling stereo components. You will be paid by piecework. What questions should you ask to find out more about how much you will be earning?

## Develop

### Example 1

Anne-Marie works part time at a large sports facility selling programs. She is paid by piecework. She earns $10 per game plus 20¢ for each program she sells. Find Anne-Marie's total earnings when she sells 93 programs at one game.

### Solution

*Step 1* Calculate her piecework earnings.

20¢ = $0.20

93 programs at 20¢ per program $= 93 \times 0.20$
$= 18.60$

*Step 2* Calculate her total earnings.

$10 plus $18.60 piecework earnings $= 10 + 18.60$
$= 28.60$

Anne-Marie's total earnings are $28.60.

### Example 2

Ethan builds wooden pallets. On Monday he builds 300 pallets and is paid 35¢ per pallet.

**a)** What are his earnings for Monday?

**b)** If he builds 300 each day from Tuesday to Friday as well, what would his earnings be for the week?

**c)** Ethan works 50 weeks a year. If he makes the same number of pallets each week, how much will he earn in a year?

### Solution

**a)** 35¢ = \$0.35

300 pallets at \$0.35 = $300 \times 0.35$
= 105

Ethan earns \$105 on Monday.

**b)** Monday plus Tuesday to Friday is 5 days in all.

5 days at \$105 per day = $5 \times 105$
= 525

Ethan would earn \$525 for the week.

**c)** 50 weeks at \$525 per week = $50 \times 525$
= 26 250

Ethan would earn \$26 250 for the year.

## Practise

**1.** Mark works at a rock concert selling programs. He is paid \$15 for the night plus 20¢ for each program that he sells. He sells 150 programs. How much does he earn working at the rock concert?

**2.** Orchards employ workers to prune trees. They pay the workers from \$2 to \$3 per tree for this service.

**a)** During Abdulmalik's first week of work, he prunes 150 trees at \$2 per tree. What are his earnings for the week?

**b)** Milan is experienced at pruning trees. He prunes 300 trees at \$2.50 per tree in one week. What are his earnings for the week?

**3.** Why might the rate given in question 2 range from \$2 to \$3 per tree?

**4.** Why do you think that some orchards often hesitate to pay fruit pickers by piecework?

**5.** Worm pickers are paid from \$20 to \$30 per 1000 worms picked. The rate varies from evening to evening.

**a)** One evening Paige is paid \$28 per 1000 worms picked. She picks 5000 worms. What are her earnings for the evening?

**b)** Arvinder earns \$168 the same evening. How many worms did he pick?

**6.** Why might the rate given in question 5 range from \$20 to \$30 per 1000 worms picked?

**7.** How does piecework benefit the worker and the employer?

# 1.4 Hourly Rate and Overtime Rate

Another way to be paid for work done is by an **hourly rate**. Workers are paid for each hour they work. When more than 44 hours a week are worked, Ontario's *Employment Standards Act* states that most workers are entitled to be paid **overtime** for the extra hours. The **overtime rate** must be at least "time and a half," or 1.5 times the regular hourly rate. Some jobs start to pay overtime for fewer hours than 44. Some jobs pay more than time and a half.

## Develop

### Example 1

Alison has a job with the city maintenance crew. She is paid an hourly rate of \$12.50. She usually works 8 hours a day, 5 days a week.

**a)** Find her **regular earnings** for one week.

**b)** If Alison works more than 40 hours a week, she is paid overtime for the extra hours. The overtime rate is time and a half, or 1.5 times the regular hourly rate. One week Alison worked 5 hours overtime. Find her total earnings for that week.

### Solution 

**a)** *Step 1* Calculate the number of **regular hours** she works in a week.
8 hours a day for 5 days $= 8 \times 5$
$= 40$

*Step 2* Calculate her regular earnings for the week.
40 hours a week at an hourly rate of \$12.50 $= 40 \times 12.50$
$= 500$

Alison's regular earnings for one week are \$500.

**b)** *Step 1* Calculate her overtime hourly rate.
1.5 times the regular hourly rate of \$12.50 $= 1.5 \times 12.50$
$= 18.75$

*Step 2* Calculate her overtime earnings.
5 hours overtime at an overtime hourly rate of \$18.75 $= 5 \times 18.75$
$= 93.75$

*Step 3* Calculate her total earnings for that week.
regular earnings plus overtime earnings $= 500 + 93.75$
$= 593.75$

Alison's total earnings for that week are \$593.75.

## Example 2

Dino is paid $9.50 per hour for 40 hours a week. If he works more than 40 hours a week, he is paid time and a half. His friend Len has a different job. Len is paid $9.20 per hour for 37.5 hours a week. If Len works more than 37.5 hours a week, he is paid time and a half. If they both work 45 hours one week, who is paid more?

### Solution

**Method 1 Using a Calculator**

Dino

*Step 1* Calculate his regular earnings.
40 hours at an hourly rate of $9.50 $= 40 \times 9.50$
$= 380$

*Step 2* Calculate his overtime hours.
45 hours of work less 40 regular hours $= 45 - 40$
$= 5$

*Step 3* Calculate his overtime hourly rate.
1.5 times the regular hourly rate of $9.50 $= 1.5 \times 9.50$
$= 14.25$

*Step 4* Calculate his overtime earnings.
5 hours at an overtime hourly rate of $14.25 $= 5 \times 14.25$
$= 71.25$

*Step 5* Calculate his total earnings.
regular earnings plus overtime earnings $= 380 + 71.25$
$= 421.25$

**Method 2 Using a Spreadsheet**

Find regular earnings, overtime hours, overtime rate, overtime earnings, and total weekly earnings.

| | A | B | C | D | E | F | G | H | I |
|---|---|---|---|---|---|---|---|---|---|
| 1 | Name | Hours Worked | Regular Hourly Rate ($) | Regular Hours | Regular Earnings ($) | Overtime Hours | Overtime Rate ($) | Overtime Earnings ($) | Total Weekly Earnings ($) |
| 2 | Dino | 45 | 9.50 | 40 | **380.00** | **5** | **14.25** | **71.25** | **421.25** |
| 3 | Len | 45 | 9.20 | 37.5 | | | | | |

Complete the work for Len. Then, compare his total weekly earnings to Dino's.

## Practise

1. a) Why would workers be paid time and a half for overtime?
   b) What does "time and a half" mean?
   c) What would "double time" mean?

2. Pierre works full time and is paid an hourly rate of $12.50 for 35 hours a week. He is paid for 52 weeks of the year. What are his total earnings for one year?

3. Jim stocks shelves at a grocery store. He earns $8.60 per hour for 37.5 hours each week. One week, a large shipment arrives late and Jim is asked to work overtime at 1.5 times his regular rate. He works 4.5 hours of overtime. What are his total earnings for the week?

4. The overtime rate is 1.5 times the regular hourly rate.
   a) Explain how the amounts given in cells E2 to I2 were calculated.
   AT b) FILL DOWN Columns E to I, Rows 2 to 6. What are the total weekly earnings for the others?

| | A | B | C | D | E | F | G | H | I |
|---|---|---|---|---|---|---|---|---|---|
| 1 | Name | Hours Worked | Regular Hourly Rate ($) | Regular Hours | Regular Earnings ($) | Overtime Hours | Overtime Rate ($) | Overtime Earnings ($) | Total Weekly Earnings ($) |
| 2 | Ellen | 44 | 8.50 | 44 | 374.00 | 0 | 12.75 | 0.00 | 374.00 |
| 3 | George | 48 | 7.50 | 44 | | | | | |
| 4 | Chris | 47 | 10.00 | 44 | | | | | |
| 5 | Hanya | 52 | 12.50 | 44 | | | | | |
| 6 | Joel | 49 | 9.50 | 44 | | | | | |

5. Steve earns an annual salary of $36 000. His brother Andrew earns $18.25 per hour and is paid for 35 hours a week for 52 weeks a year.
   a) How much more than his brother does Steve earn in a year?
   b) Apply what you have learned to determine the number of overtime hours at time and a half that Andrew would need to work to earn more than his brother earns in a year.

### Skills Check — Percent of a Number

Use a calculator to find each percent.

a) 20% of 25 b) 15% of 50 c) 12% of 40
d) 7% of 32 e) 8% of 41 f) 5% of 28

# 1.5 Career Focus: Restaurant Server

Sunil and Moira apply for jobs that they saw advertised at a busy restaurant in the shopping mall.

The ad indicated
- an hourly rate of $7.10 for greeters
- an hourly rate of $6.85 plus tips for servers
- the importance of cleanliness

Requirements for both positions include
- an outgoing personality
- reliable work habits
- punctuality
- an excellent attendance record
- reliable organizational skills

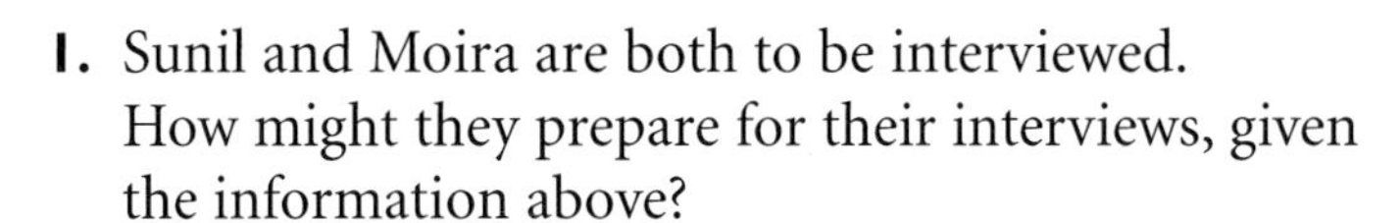

1. Sunil and Moira are both to be interviewed. How might they prepare for their interviews, given the information above?

2. The interviewer tells them that successful candidates will be contacted between 5 p.m. and 6 p.m. the next day.
   **a)** How should Moira and Sunil arrange their schedules the next day?
   **b)** What message would it send to the potential employer, if they could not be reached between 5 p.m. and 6 p.m.?

3. While being interviewed, Moira and Sunil were told that
   - servers and greeters work 6-hour shifts
   - servers usually serve $100 worth of food and beverages per hour
   - servers could expect a 10% to 15% tip on all food and beverage sales

   Immediately after their interviews, Moira and Sunil estimate their potential earnings.

   **a)** What would a greeter earn working a 6-hour shift?
   **b)** What would a greeter earn working five 6-hour shifts?
   **c)** Based on 10% tips on $100 worth served each hour, what would a server earn working a 6-hour shift?
   **d)** Based on 10% tips on $100 worth served each hour, what would a server earn working five 6-hour shifts?

**4.** During their interviews, both Moira and Sunil demonstrated that they met the requirements for both types of jobs. How might they have demonstrated their suitability for the jobs during their interviews?

**5.** **a)** Do you meet the requirements for the jobs of server and greeter? Explain.

**b)** Would you have accepted either of the jobs? Why or why not?

**MENU**

| **Starters** | |
|---|---|
| Soup of the Day | $3.95 |
| Bruschetta | $3.55 |
| Green salad | $3.50 |
| Caesar salad | $4.25 |
| dinner size | $5.25 |
| **Entrees** | |
| Hamburger with fries | $6.95 |
| Chicken fajitas | $8.95 |
| Personal pizza | $7.25 |
| Pasta of the Day | $7.95 |
| **Desserts** | |
| Chocolate mousse | $3.25 |
| Apple pie | $2.99 |
| Fresh fruit cup | $2.75 |
| Sundae | $3.10 |
| **Beverages** | |
| Milk | $1.25 |
| Juice | $1.75 |
| Milkshakes | $2.50 |
| Tea/Coffee | $0.95 |
| Soft drinks | $1.50 |

**6.** Sunil and Moira both accept jobs as servers. When taking customers' food orders, they assist customers with their selections. They must know the daily features. What two items on the menu change daily?

**7.** Servers are instructed to ask customers if they want to order something from each of the four menu categories. One customer orders a green salad, a hamburger, and chocolate mousse. What would be an appropriate question for this customer?

**8.** A customer asks a server to suggest a meal that he might enjoy for under $12. What would be a suitable recommendation?

**9.** A customer brings to a server's attention that she has a severe allergy to eggs and peanuts. What should the server do to ensure this customer's safety?

# 1.6 Commission

**Commission**, another way to be paid, is calculated as a percent of sales made. **Straight commission** means that earnings are based on commission alone. **Salary plus commission** means that commission is combined with a salary.

## Explore

You have been offered a part-time job selling sporting goods. You can be paid minimum wage, salary plus commission, or straight commission. What questions should you ask to help you decide which way to be paid?

## Develop

### Example 1

Fatemeh works in a shoe store. She earns a salary of \$240 a week plus 6% commission. One week she sold \$3000 worth of shoes. Find her total earnings for the week.

### Solution

*Step 1* Calculate her commission.

$$\begin{aligned} \text{6\% commission on \$3000 worth of shoes} &= \text{6\% of \$3000} \\ &= 0.06 \times 3000 \\ &= 180 \end{aligned}$$

*Step 2* Find her total earnings.

$$\begin{aligned} \text{\$180 commission plus \$240 salary} &= 180 + 240 \\ &= 360 \end{aligned}$$

Fatemeh's total earnings are \$360 for the week.

## Example 2

Martha works part time selling cosmetics. She is paid 30% of her sales. Her sales for each work day of one week are given. Calculate her earnings for the week.

| Mon. | Tues. | Wed. | Thurs. | Fri. |
|---|---|---|---|---|
| $87.00 | $76.50 | $120.60 | $60.00 | $85.90 |

### Solution

**Method 1 Using a Calculator**

*Step 1* Total the sales.

$87.00 + 76.50 + 120.60 + 60.00 + 85.90 = 430$

*Step 2* Calculate her commission.

30% commission on \$430 total sales $= 0.30 \times 430$
$= 129$

**Method 2 Using a Spreadsheet**

Total the sales and determine 30% of the total.

| | A | B | C | D | E | F | G |
|---|---|---|---|---|---|---|---|
| 1 | Daily Sales ($) | | | | | Total | Total |
| 2 | Mon. | Tues. | Wed. | Thurs. | Fri. | Sales ($) | Earnings ($) |
| 3 | 87.00 | 76.50 | 120.60 | 60.00 | 85.90 | **430.00** | **129.00** |

Martha's total earnings for the week are $129.

## Practise

**1.** Eddie receives a monthly salary of $800 and 5% commission on his total sales for a month, $31 000. Calculate his total earnings for the month.

**2.** Mohammed sells furniture and is paid 6% of his sales. His sales for one week are given. Calculate his earnings for the week.

| Tues. | Wed. | Thurs. | Fri. | Sat. |
|---|---|---|---|---|
| $1600 | $1250 | $1390 | $1840 | $2300 |

**3.** The rate of commission is 25% of total sales.

**a)** Explain how the amounts given in cells G3 and H3 were calculated.

AT **b)** FILL DOWN Columns G and H, Rows 3 to 7. What are the total earnings for the others?

| | A | B | C | D | E | F | G | H |
|---|---|---|---|---|---|---|---|---|
| 1 | Name | Daily Sales ($) | | | | | Total | Total |
| 2 | | Mon. | Tues. | Wed. | Thurs. | Fri. | Sales ($) | Earnings ($) |
| 3 | Anne | 425 | 422 | 431 | 370 | 429 | 2077.00 | 519.25 |
| 4 | John | 434 | 424 | 430 | 422 | 432 | | |
| 5 | Hans | 425 | 429 | 433 | 439 | 434 | | |
| 6 | Adam | 419 | 422 | 394 | 420 | 423 | | |
| 7 | Rita | 380 | 428 | 425 | 398 | 417 | | |

**4.** Joyce sells cars. She earns 35% commission on the profit of each sale. One week she sells two cars. The profits on the cars are $800 and $1300. What are her earnings for the week?

**5.** Mark sells magazine subscriptions by telephone. He is paid $256 a week and receives a commission for each subscription he sells. In one week he sold the subscriptions shown.

**a)** Explain how the amount given in cell E2 was calculated.

AT **b)** FILL DOWN Column E, Rows 2 to 6, and SUM cell E7. What is Mark's total weekly commission?

**c)** Explain how the amount that appears in cell E7 was calculated.

**d)** What are Mark's total weekly earnings?

| | A | B | C | D | E |
|---|---|---|---|---|---|
| 1 | Magazine | Price ($) | Number Sold | Rate of Commission (%) | Commission ($) |
| 2 | McLaren's | 38.50 | 12 | 4 | 18.48 |
| 3 | Mechanic's Monthly | 28.50 | 15 | 5 | |
| 4 | Popular People | 46.50 | 10 | 7 | |
| 5 | TV to Watch | 78.50 | 8 | 3 | |
| 6 | Living Canadian | 36.50 | 10 | 4 | |
| 7 | Total Weekly Commission | | | | |

**6.** Consider straight commission and salary plus commission.

**a)** What are some advantages and disadvantages of each method of payment for the salesperson?

**b)** What personal characteristics are needed for sales jobs that pay commissions?

# 1.7 Step Commission

## Develop

Shamir sells clothing. He earns a monthly salary of $900, 5% commission on the first $20 000 in sales he makes in a month, and 8% on sales over $20 000. His sales for each month are shown in the spreadsheet below.

**1.** Refer to the spreadsheet below.

**a)** For which month did Shamir have the greatest sales? Why do you think that was?

**b)** For which month did he have the fewest sales? Why might that have been?

**c)** For how many months does Shamir receive only a 5% commission on his sales?

**2.** Explain why each statement is true.

**a)** In March Shamir receives 8% commission on sales of $5175.

**b)** Shamir's earnings in March are made up of these three amounts:
$900 5% of $20 000 8% of $5175

AT **3.** Complete the spreadsheet to calculate Shamir's total earnings each month and his total earnings for the year by following the steps on the next page.

| | A | B | C | D | E | F | G |
|---|---|---|---|---|---|---|---|
| 1 | Month | Salary ($) | Sales ($) | 5% Commission on Sales Less Than or Equal to $20000 ($) | Sales Over $20000 ($) | 8% Commission on Sales Greater Than $20000 ($) | Total Earnings ($) |
| 2 | Jan. | 900 | 18245 | 912.25 | 0.00 | 0.00 | 1812.25 |
| 3 | Feb. | | 19450 | | | | |
| 4 | Mar. | | 25175 | | | | |
| 5 | Apr. | | 24199 | | | | |
| 6 | May | | 28317 | | | | |
| 7 | June | | 14259 | | | | |
| 8 | July | | 22082 | | | | |
| 9 | Aug. | | 27635 | | | | |
| 10 | Sept. | | 29452 | | | | |
| 11 | Oct. | | 24674 | | | | |
| 12 | Nov. | | 29285 | | | | |
| 13 | Dec. | | 31259 | | | | |
| 14 | Total | | | | | | |

a) FILL DOWN Column B, Rows 2 to 13. What do you get for the salary for each month? Is that what you would expect?

b) FILL DOWN Column D, Rows 2 to 13. Which months are less than $1000? Explain why.

c) FILL DOWN Column E, Rows 2 to 13. Which months are $0.00?

d) FILL DOWN Column F, Rows 2 to 13. Which months are $0.00? Explain why.

e) FILL DOWN Column G, Rows 2 to 13. In which month did Shamir have the greatest earnings?

f) SUM Column B. What is Shamir's salary for the year? Is that what you would expect? Explain.

g) SUM Column G. What are his total earnings for the year?

**4.** Explain how the amounts for May in cells D6, E6, F6, and G6 and the amount in cell G14 were calculated.

## Practise

**5.** Linda sells cosmetics. She earns a monthly salary of $1000, 12% commission on the first $5000 in sales she makes in a month, and 15% on sales over $5000.

a) Explain how the amounts given in cells D2, E2, F2, and G2 were calculated.

AT b) Complete the spreadsheet to calculate Linda's total earnings each month and her total earnings for the year.

| | A | B | C | D | E | F | G |
|---|---|---|---|---|---|---|---|
| 1 | Month | Salary ($) | Sales ($) | 12% Commission on Sales Less Than or Equal to $5000 ($) | Sales Over $5000 ($) | 15% Commission on Sales Greater Than $5000 ($) | Total Earnings ($) |
| 2 | Jan. | 1000 | 4825 | 579.00 | 0.00 | 0.00 | 1579.00 |
| 3 | Feb. | | 2765 | | | | |
| 4 | Mar. | | 5592 | | | | |
| 5 | Apr. | | 5339 | | | | |
| 6 | May | | 5488 | | | | |
| 7 | June | | 5793 | | | | |
| 8 | July | | 5504 | | | | |
| 9 | Aug. | | 5206 | | | | |
| 10 | Sept. | | 5712 | | | | |
| 11 | Oct. | | 5683 | | | | |
| 12 | Nov. | | 5290 | | | | |
| 13 | Dec. | | 6183 | | | | |
| 14 | Total | | | | | | |

# 1.8 Putting It All Together: Finding a Job

Job application forms typically request information about the following:

- personal data
- education
- employment history
- job applied for
- availability
- interests and experience

Under **Personal Data**, you are asked information, such as your name, address, and phone number.

Under **Education**, you are usually asked what level of education you have completed.

Under **Employment History**, you are typically asked about your last two or three jobs.

**1.** Complete the following portions of the job application form that your teacher gives you:

- Personal Data
- Education
- Employment History

**2.** You are frequently asked to identify any special interests, skills, training, and experience that would help you in the job for which you are applying. List such features about yourself that could be useful in any job you might consider.

**3. a)** Use ads from newspapers or the Internet to find two or three interesting, full-time jobs for which you are or could become qualified.

**b)** Give reasons for your choices.

Access to Web sites for jobs can be gained through the *Mathematics for Everyday Life 11* page of irwinpublishing.com/students.

WS **4.** For each job you found in question 3, identify the following.

**a)** the duties
**b)** the location
**c)** the method of payment (e.g., salary, hourly rate, commission, piecework)
**d)** the required personal characteristics
**e)** the required education
**f)** the required experience

WS **5.** Consider the location of each job.

**a)** Would you need to relocate?
**b)** How would you get to and from work?

**6.** For each job, is there information about amount of pay and hours to be worked? If not, discuss with your teacher what might be typical for such a job.

WS **7.** For each job, determine

**a)** the number of hours you would expect to work each week
**b)** how much you would expect to earn in a week, a month, a year

**8.** Determine which job suits you best, in light of

- your answers to questions 4 to 7
- the fact that your earnings will affect the type of housing, transportation, investing, travelling, and so on that you will be able to afford

Explain your choice.

**9.** For the job you selected, answer the following.

**a)** To whom should you apply?
**b)** How should the person be contacted?

**10. a)** Write a brief letter, fax, or e-mail message explaining that you are applying for the job you selected.

WS **b)** Complete the Position Applied For and Availability sections of the job application. Use your answer to question 2 to help you with the last part.

***You got the job! In the next chapters, you will use the information about your earnings from this job to determine payroll deductions, income tax payable, savings, investments, vacation plans, and a car purchase.***

# 1.9 Chapter Review

1. **a)** Name four different methods of payment.
   **b)** Which allows for being paid overtime?
   **c)** Which is most often associated with sales positions?

2. Chandra works as a groundskeeper for a hospital. Her salary is $26 780 per year.
   **a)** How much are her earnings per week?
   **b)** How much are her earnings per month?

3. Claudia works at a trade fair signing up customers. She is paid $50 for the day and $1 for each customer she signs up. How many customers does she need to sign up to double what she would earn if she didn't sign up any customers?

4. The overtime rate is 1.5 times the regular hourly rate.
   **a)** Explain how the amounts given in cells E2 to I2 were calculated.
   AT **b)** FILL DOWN Columns E to I, Rows 2 to 6. What are the total weekly earnings for the others?

| | A | B | C | D | E | F | G | H | I |
|---|---|---|---|---|---|---|---|---|---|
| 1 | Name | Hours Worked | Regular Hourly Rate ($) | Regular Hours | Regular Earnings ($) | Overtime Hours | Overtime Rate ($) | Overtime Earnings ($) | Total Weekly Earnings ($) |
| 2 | Mike | 45 | 7.50 | 44 | 330.00 | 1 | 11.25 | 11.25 | 341.25 |
| 3 | Isobel | 45 | 9.50 | 44 | | | | | |
| 4 | Denise | 50 | 12.60 | 44 | | | | | |
| 5 | Peter | 48 | 8.50 | 44 | | | | | |
| 6 | Sara | 47 | 7.75 | 44 | | | | | |

5. Jennifer installs and repairs telephones. Her hourly rate is $18.50. She usually works 8 hours a day, 5 days a week. She is paid double time for each hour she works on holidays or weekends. One week Jennifer works her regular hours and 5 hours on Saturday. What are her earnings for the week?

6. Gerry installs glass in automobiles.
   **a)** He used to be paid $11 an hour for 35 hours a week. What were his weekly earnings?
   **b)** He is now paid $20 per unit of glass installed. He installs 21 units this week. What are his weekly earnings?
   **c)** How does the change to payment by piecework benefit Gerry and his employer?

7. Greg clears tables at a restaurant. He is paid $7.10 per hour plus 4% of all tips collected by the servers. One week he worked 30 hours and the tips while he worked totalled $1350. Calculate Greg's total earnings for the week.

8. Calculate the earnings for each.

| | Amount of Sales | Rate of Commission |
|---|---|---|
| a) | $500 | 4% |
| b) | $2500 | 10% |
| c) | $750 | 25% |
| d) | $1250 | 28% |

9. Kim sells furniture. She receives a monthly salary of $1000 and earns 3% commission on her sales. Her sales for each week of one month are shown. Find her total monthly earnings.

| Week 1 | Week 2 | Week 3 | Week 4 |
|---|---|---|---|
| $12 225 | $7229 | $8445 | $9740 |

10. Barbara sells hairdressing supplies. She earns a weekly salary of $300, 6% commission on the first $1000 in sales she makes each week, and 10% on sales over $1000.

a) Explain how the amounts given in cells D2, E2, F2, and G2 were calculated.

AT b) FILL DOWN Columns B and D to G, Rows 2 to 5. What are Barbara's total earnings for the other weeks?

| | A | B | C | D | E | F | G |
|---|---|---|---|---|---|---|---|
| 1 | Week | Salary ($) | Sales ($) | 6% Commission on Sales Less Than or Equal to $1000 ($) | Sales Over $1000 ($) | 10% Commission on Sales Greater Than $1000 ($) | Total Earnings ($) |
| 2 | 1 | 300 | 1250 | 60.00 | 250.00 | 25.00 | 385.00 |
| 3 | 2 | | 980 | | | | |
| 4 | 3 | | 1100 | | | | |
| 5 | 4 | | 1400 | | | | |

11. For 35 hours a week and 52 weeks of the year, what hourly rate is equivalent to an annual salary of $32 000?

A 12. You would like to earn at least $500 a week. You have three part-time job offers.

- One pays $10 an hour, 24 hours a week.
- Another pays $7 an hour, 18 hours a week, plus 5% commission on sales of about $2500.
- The third pays $6.85 an hour plus about $10 each hour in tips for 16 hours a week.

What could you do to achieve your goal? Justify your decision.

CHAPTER

# 2 Deductions and Expenses

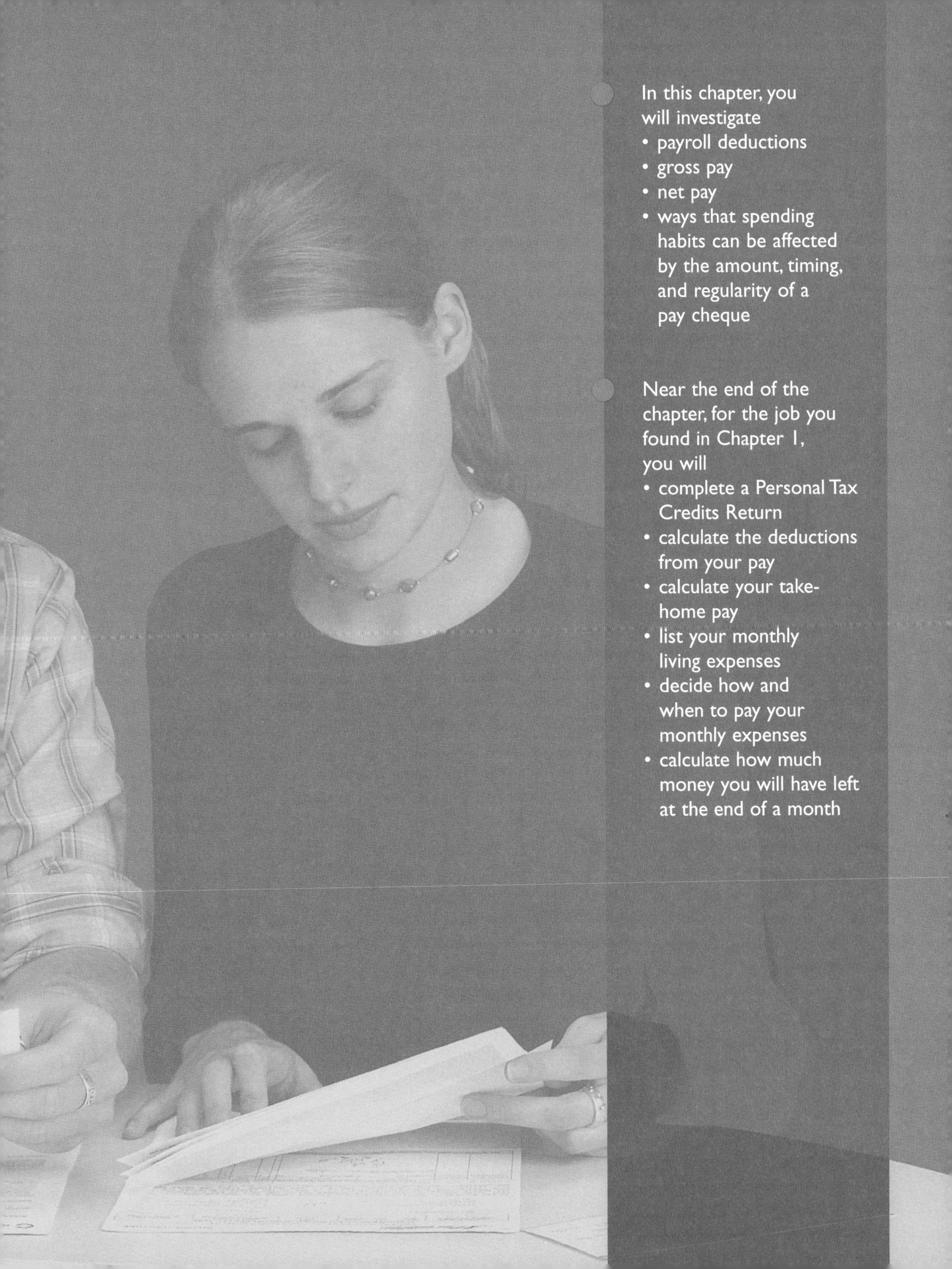

In this chapter, you will investigate

- payroll deductions
- gross pay
- net pay
- ways that spending habits can be affected by the amount, timing, and regularity of a pay cheque

Near the end of the chapter, for the job you found in Chapter 1, you will

- complete a Personal Tax Credits Return
- calculate the deductions from your pay
- calculate your take-home pay
- list your monthly living expenses
- decide how and when to pay your monthly expenses
- calculate how much money you will have left at the end of a month

# 2.1 Standard Deductions

## Explore

Kevin is a new employee at MmmPizza. Each week he works 37.5 hours at $7 per hour. His pay cheque is less than $225. Why?

## Develop

**1.** Kevin, like all new employees, completed a TD1 form, a **Personal Tax Credits Return**. Use the portion of the Personal Tax Credits Return shown below.

**a)** What is Kevin's full name?
**b)** In what month was he born?
**c)** What is his SIN (Social Insurance Number)?

**2.** Do you have a SIN? When did you get it?

Canada Customs and Revenue Agency / Agence des douanes et du revenu du Canada

**2001 PERSONAL TAX CREDITS RETURN** TD1

Complete this TD1 form if you have a new employer or payer and you will receive salary, wages, commissions, pensions, Employment Insurance benefits, or any other remuneration. Be sure to sign and date it on the back page and give it to your employer or payer who will use it to determine the amount of your payroll tax deductions.

If you do not complete a TD1 form, your new employer or payer will deduct taxes after allowing the basic personal amount **only**.

You **do not** have to complete a new TD1 form every year unless there is a change in your personal tax credit amounts. Complete a new TD1 form no later than seven days after the change.

You can get the forms and publications mentioned on this form from our Internet site at **www.ccra-adrc.gc.ca** or by calling 1-800-959-2221.

| Last name | First name and initial(s) | Date of birth (YYYY/MM/DD) | Employee number |
|---|---|---|---|
| Wilmont | Kevin J. | 1984/06/14 | 129 |
| Address including postal code | | For non-residents only – Country of permanent residence | Social insurance number |
| 432 Rosewood Court Ourtown, ON M4L 2S5 | | | 3 4 3 5 3 5 4 4 4 |

**3.** How Kevin completes the rest of the form tells his employer how much money to take off his pay for three standard government **deductions**. These deductions are subtracted from the amount of money earned, that is, **gross pay**, or **gross income**.

**Net pay**, or **net income**, is the gross pay minus the total of all the deductions. It is often called **take-home pay**.

**a)** Kevin works 37.5 hours a week at $7 per hour. What is his gross pay?
**b)** Kevin's deductions are $37.80. What is his net pay?

4. **Income tax** is one category of government deduction. It is a portion of earnings collected by the federal and provincial governments to pay for services supplied to us by the governments. The greater a worker's earnings, the greater the deduction.
   **a)** What are five services that taxes pay for?
   **b)** How could you determine which level of government supplies a given service?

5. Kevin's federal income tax deduction is $17 and his provincial income tax deduction is $6.50 each week.
   **a)** How much income tax is deducted each week?
   **b)** How much income tax would be deducted in a year?

6. **EI (Employment Insurance)** is another government deduction. It is a fund into which employees and employers pay. If workers lose their jobs through no fault of their own and have worked at least 931 hours, they are eligible to collect monthly EI payments for a period of time. The greater a worker's earnings, the greater the deduction and EI payments, if collected.

   Kevin's deduction for EI is $5.91 each week. How much would be deducted in a year for EI?

7. **CPP (Canada Pension Plan)** is the third government deduction. It is a pool of money to which employees and employers contribute. It is used to pay monthly pensions to workers upon retirement. The greater a worker's earnings, the greater the contribution and the pension received during retirement.

   Kevin's deduction for CPP is \$8.39 a week. How much would be deducted in a year for CPP?

8. Use the information about Kevin's deductions from questions 5, 6, and 7.

   **a)** Which deduction—income tax, EI, or CPP—is the greatest?
   **b)** Why do you think the government deducts more for that than for the other two?

## Skills Check — One Number as a Percent of Another

Use a calculator, and round to the nearest whole number.

**a)** the number of boys in your class as a percent of the number of students in the class
**b)** the number of people in the room wearing glasses as a percent of the number of people in the room
**c)** the number of math classes you have today as a percent of the number of classes you have today
**d)** the number of hours you spend at school as a percent of the number of hours in a day

Use a calculator, and round to the nearest tenth.

**e)** Kevin's CPP deduction, \$8.39, as a percent of his gross pay, \$262.50

Use a calculator, and round to the nearest hundredth.

**f)** Kevin's EI deduction, \$5.91, as a percent of his gross pay, \$262.50

# 2.2 Other Deductions

Some jobs require deductions in addition to EI, CPP, and income tax.

Deductions may also be made for
- a company pension plan
- a health plan
- a life insurance plan
- a disability insurance plan
- union dues
- a payroll savings plan
- charitable donations

Sometimes, the employer, as well as the employee, contributes to pension plans or health plans. The portion that the employer contributes is considered a **benefit**.

If a company **pension plan** exists, all employees contribute every pay period. If they are still working for the company up till retirement, they get a company pension as well as Canada Pension.

A company **health plan** would cover some medical expenses not covered by OHIP (Ontario Health Insurance Plan). These include
- prescription glasses and contacts
- dental checkups and dental work
- semi-private or private rooms in hospitals

If a company **life insurance plan** or **disability insurance plan** is offered, employees can obtain insurance at lower group rates.

**Union dues** are deducted from pay if a job requires membership in a union. Unions negotiate hourly rates or salaries, benefits, and working conditions with employers.

Employees may have the option of taking part in a **payroll savings plan**. For example, they might purchase Canada Savings Bonds, where some money to pay for the bonds is deducted from each pay.

Employees may have the option of making **donations to charities**. For example, they might make donations to the United Way in their community, where some money for the donations is deducted from each pay.

## Develop

### Example

Karen earns $380 per week working for a telemarketing company.
The deductions from her weekly pay are

| | |
|---|---|
| EI | $8.55 |
| CPP | $13.45 |
| Income tax | $44.15 |
| Company pension plan | $9.50 |
| Health plan | $10.40 |

Determine Karen's

**a)** gross pay **b)** total deductions **c)** net pay

### Solution

**a)** Karen's gross pay is what she earns, $380.

**b)** total deductions $= 8.55 + 13.45 + 44.15 + 9.50 + 10.40$
$= 86.05$

Her total deductions are $86.05.

**c)** net pay = gross pay − total deductions
$= 380 - 86.05$
$= 293.95$

Her net pay is $293.95.

## Practise

**1.** List possible deductions from a pay cheque.

**2.** Is a deduction or a benefit described in each situation?

**a)** Jason has a company car.
**b)** Rama's company pays for a basic health plan.
**c)** Liam is buying company stocks through a payroll plan.

**3.** Joy's gross income is $427. Here are her deductions:

| | |
|---|---|
| EI | $9.61 |
| CPP | $15.47 |
| Income tax | $57.55 |
| Company pension plan | $10.40 |

Determine her total deductions and her net income.

**4.** Clement's gross pay is $372. Here are his deductions:

| | |
|---|---|
| EI | $8.37 |
| CPP | $13.10 |
| Income tax | $42.60 |
| Company pension plan | $9.30 |
| Union dues | $6.20 |

Determine his total deductions and his net pay.

**5.** Determine the missing information for each person.

| | Name | Gross Pay | Deductions | Net Pay |
|---|---|---|---|---|
| **a)** | Win | $257.20 | $42.49 | |
| **b)** | Margaret | | $73.11 | $311.96 |
| **c)** | Mark | $429.00 | | $347.30 |
| **d)** | Clare | | $109.50 | $388.50 |

**6. a)** How does a higher gross income affect deductions?

**b)** Sam wants to know how much he can afford to pay for rent. Should he consider his gross income or his net income? Explain.

**7.** Lim earns $7.98 per hour. He works 36 hours a week. Here are his weekly deductions:

| | |
|---|---|
| EI | $6.46 |
| CPP | $9.47 |
| Income tax | $28.55 |
| Company pension plan | $7.18 |
| Savings plan | $25.00 |

Determine his weekly

**a)** gross pay **b)** total deductions **c)** net pay

**8.** Zoe earns $11.50 per hour. She works 35 hours a week. Here are her weekly deductions:

| | |
|---|---|
| EI | $9.06 |
| CPP | $14.41 |
| Income tax | $49.10 |
| Company pension plan | $10.77 |
| Health plan | $4.85 |

Determine her weekly

**a)** gross pay **b)** total deductions **c)** net pay

**9.** Gopal works 35 hours a week. His net income is $348 and his deductions are $89.50. Apply what you have learned to determine his hourly rate.

# 2.3 Career Focus: Amusement Park Worker

Kendrick has a seasonal job at an amusement park. He works from early April to the end of October. His job focuses on one ride, which he

- inspects daily
- maintains, including doing minor repairs
- operates and sells tickets for

Before the park opens in May and after it closes mid-October, he does general park maintenance.

Kendrick earns $9.25 per hour and works a 6.5-hour shift six days each week. His day off varies.

**1. a)** What is Kendrick's gross weekly pay?

**b)** Approximate the number of weeks he works and estimate his gross annual pay.

**2.** Kendrick's weekly deductions are

| | |
|---|---|
| Income tax | $43.40 |
| EI | $8.12 |
| CPP | $12.62 |

**a)** How much is Kendrick's take-home pay?

**b)** What percent of his gross pay is his take-home pay, rounded to the nearest whole number?

**3.** Kendrick is eligible for EI benefits when he is not working. Do you think his EI payments would be less than, the same as, or more than his take-home pay? Why do you think so? How could you find out?

**4.** Daily inspection and maintenance of the ride includes

- checking the functioning of several parts
- tightening nuts and bolts
- recording the time and date of the inspection in a log book
- reading the repair manual, safety notices, and repair notices from the manufacturer
- filing notices in the repair manual
- deciding if the ride will operate safely each day

What skills are important for the part of Kendrick's job just described? Why do you think so?

## CONTROL SHEET SUMMARY

Date July 5 Prepared by Kendrick Black

Location Circling the World On duty with Pamela Grey

**Ticket Start and Stop Numbers**

Start number 57 436 Stop number 57 887 Number of tickets sold 451

**A: TICKET SUMMARY**

| Tickets × Charge | Total Sales |
|---|---|
| 451 × 0.75 = | ________ |

**B: CASH SUMMARY**

| | Bills | Amount | | Coins | Amount |
|---|---|---|---|---|---|
| 6 × | 5.00 = | ________ | 5 × | 0.01 = | ________ |
| 5 × | 10.00 = | ________ | 4 × | 0.05 = | ________ |
| 4 × | 20.00 = | ________ | 5 × | 0.10 = | ________ |
| 1 × | 50.00 = | ________ | 18 × | 0.25 = | ________ |
| 0 × | 100.00 = | ________ | 49 × | 1.00 = | ________ |
| | | | 37 × | 2.00 = | ________ |
| Total Bills | | ________ | Total Coins | | ________ |

| | |
|---|---|
| Total Bills | ________ |
| Total Coins | ________ |
| Total Cash | ________ |

**5.** To operate the ride and sell tickets, Kendrick is teamed with another employee. They take turns doing each task. At the end of the day, Kendrick completes a Control Sheet Summary, like the one shown.

**a)** Who worked with Kendrick on July 5?

**b)** How many tickets were sold for his ride that day?

**c)** How much does each ticket cost?

**6.** Explain how the total number of tickets sold is calculated.

WS **7. a)** Complete the Ticket Summary and Cash Summary calculations for Kendrick.

**b)** Are Total Sales and Total Cash equal? Check your work if they are not.

**8.** What skills are important for the part of Kendrick's job that involves operating the ride and selling tickets? Why do you think so?

***Kendrick is now studying small engine repair in the evenings so that he can work at the park all year, overhauling equipment in the winter months.***

## 2.4 Living Expenses

### Explore

You have a job and get a pay cheque every two weeks. What should you think about when deciding what to do with the money you earn?

### Develop

WS

**1.** Some expenses occur regularly. Buying groceries weekly and paying a phone bill monthly are examples. Other expenses, such as dental checkups, arise less often. Still others, such as buying clothes, come up irregularly.

**a)** Create a list of all such expenses for a young working adult without children.
**b)** Identify each expense as essential or non-essential.
**c)** Research the amount of each expense.

**2.** Express all expenses from question 1 as monthly amounts for budgeting purposes. For example,
- if buying groceries costs \$50 each week, then the monthly amount for budgeting is $\$50 \times 4$, or \$200
- if a dental checkup every six months is \$80, the monthly amount for budgeting is $\$80 \div 6$, or \$13.33

**3.** Determine the net monthly income that would be needed to pay all essential expenses and

**a)** no other expenses
**b)** all identified non-essential expenses
**c)** selected non-essential expenses (specify which)
**d)** save \$100 a month
**e)** selected non-essential expenses (specify which), and save \$150 a month

**4.** Present your results from questions 1 to 3 to the class.

# 2.5 Comparing Expenses

Many factors affect living expenses. For example, rent can be higher in one part of the province than another and even in one part of a city than another.

## Develop

**1. a)** Research the cost of renting different types of accommodations in your community.

**b)** Discuss the advantages and disadvantages of sharing accommodations.

**2.** Explain how each benefit or job requirement listed affects a person's expenses.

**a)** wearing uniforms that are paid for, but must be dry cleaned

**b)** having a company health plan

**c)** having a company car

**3.** How does each person's situation affect his or her expenses? Why?

**a)** Cameron's rent includes heat, electricity, and water. Sharon's rent does not.

**b)** Jackson and Felica both wear glasses, but Jackson does not have a company health plan and Felica does.

**c)** Joel can walk to work. Trevor lives 12 km from his job.

**d)** Gina and her sister both spend a lot of time on the phone. However, Gina has a front-line job as a receptionist and her sister, a job in telemarketing.

**e)** Tricia and Fatima have the same job. Tricia is 22 years old and single. Fatima, 30, has a child to support.

## Practise

4. Some sources consider the following as essential expenses.
   - income taxes, CPP, and EI
   - housing
   - heath care
   - transportation
   - food
   - clothing and personal care items

   How does this list compare with the expenses that you considered essential in question 1 b) of Section 2.4?

5. Consider the following.
   - By sharing an apartment and its expenses, two friends increase their purchasing power.
   - Several people join to form a group that buys items in quantity for lower prices than they could individually. In so doing, they increase their purchasing power.
   - A person earned $10 an hour several years ago. A loaf of bread cost 39¢ then. A person earns $10 an hour today. The person today has less purchasing power than the person had several years ago.

   Discuss what you think **purchasing power** means.

6. The United Nations' Human Development Index lists the following as measures of **standard of living**.
   - wealth
   - literacy
   - access to medical care
   - employability
   - rights

   **a)** Which of these measures are influenced by education, training, and work experience? Explain.

   **b)** Which of these measures give Canadians a higher standard of living than people living in developing nations? Explain.

### Skills Check — Percent of a Number

Use a calculator to find each percent.

**a)** 5% of $150 **b)** 12% of $395 **c)** 4% of $1200
**d)** 2% of $8050 **e)** 15% of $1450 **f)** 17% of $1595

# 2.6 Paying Expenses

## Explore

Examine the following spreadsheets which show pays and expenses. How are they different? Which situation would you prefer? Explain.

| | A | B | C | D | E |
|---|---|---|---|---|---|
| 1 | Date | Item | Pay ($) | Expense ($) | Balance ($) |
| 2 | Oct. 1 | Starting balance | | | 201 |
| 3 | Oct. 1 | bus pass | | 42 | 159 |
| 4 | Oct. 4 | phone | | 25 | 134 |
| 5 | Oct. 5 | pay | 295 | | 429 |
| 6 | Oct. 6 | groceries | | 50 | 379 |
| 7 | Oct. 12 | pay | 295 | | 674 |
| 8 | Oct. 13 | groceries | | 50 | 624 |
| 9 | Oct. 15 | cable | | 22 | 602 |
| 10 | Oct. 19 | pay | 295 | | 897 |
| 11 | Oct. 20 | groceries | | 50 | 847 |
| 12 | Oct. 23 | credit card purchases | | 265 | 582 |
| 13 | Oct. 26 | pay | 295 | | 877 |
| 14 | Oct. 27 | groceries | | 50 | 827 |
| 15 | Oct. 31 | rent, including utilities | | 475 | 352 |

| | A | B | C | D | E |
|---|---|---|---|---|---|
| 1 | Date | Item | Pay ($) | Expense ($) | Balance ($) |
| 2 | Oct. 1 | Starting balance | | | 201 |
| 3 | Oct. 1 | bus pass | | 42 | 159 |
| 4 | Oct. 4 | phone | | 25 | 134 |
| 5 | Oct. 6 | groceries | | 50 | 84 |
| 6 | Oct. 13 | groceries | | 50 | 34 |
| 7 | Oct. 15 | cable | | 22 | 12 |
| 8 | Oct. 15 | pay | 1180 | | 1192 |
| 9 | Oct. 20 | groceries | | 50 | 1142 |
| 10 | Oct. 23 | credit card purchases | | 265 | 877 |
| 11 | Oct. 27 | groceries | | 50 | 827 |
| 12 | Oct. 31 | rent, including utilities | | 475 | 352 |

You have seen that gross pay is reduced by deductions and that net pay is what remains. Living expenses are paid out of net pay.

When expenses are paid depends upon factors such as
- frequency of pay periods, or how often a person is paid
- pay dates
- due dates of the expenses

Common pay periods are
- weekly—once per week
- biweekly—once every two weeks
- semimonthly—twice per month
- monthly—once per month

## Develop

**Use the first spreadsheet on page 35 to answer questions 1 to 4.**

**1.** Emelio works at Java Junction Coffee Shop. His net weekly pay is $295. He is paid every Friday. He shares an apartment with a friend. Every Saturday he spends $50 on groceries. Each of his pays and all of his expenses for October are shown in the first spreadsheet on the previous page.

Emelio's rent, due on October 31, includes utilities. What do you think "including utilities" means?

**2.** **a)** With how much money does Emelio start the month?
**b)** With how much money does he end the month?

**3.** **a)** When does Emelio have the least amount of money in October?
**b)** When does he have the most?

**4.** Emelio paid his bills when they were due.

**a)** Why do you think he avoided paying them after their due dates?
**b)** Why do you think he decided against paying them before the due dates?

**Use the second spreadsheet on page 35 to answer questions 5 and 6.**

**5.** If Emelio was paid $1180 once a month on October 15, the spreadsheet of his pays and expenses would be the second spreadsheet on page 35.

**a)** When does Emelio have the least amount of money now?
**b)** When does he have the most now?

**6.** **a)** If Emelio's starting balance had been just $12 less, what would his balance just before he received his pay have been?
**b)** If his starting balance had been $20 less, what would you suggest he do to avoid running out of money before being paid?

## Practise

**Use the following information to answer questions 7 to 11.**

Caroline's biweekly take-home pay is $996. She is paid October 1 and October 15. Her October expenses are as follows:

| | | |
|---|---|---|
| groceries | $60 | October 5 |
| credit card payment | $477 | October 12 |
| groceries | $60 | October 12 |
| car payment | $238 | October 14 |
| cable | $44 | October 15 |
| groceries | $60 | October 19 |
| car insurance | $125 | October 20 |
| phone | $36 | October 23 |
| groceries | $60 | October 27 |
| rent, including utilities | $750 | October 31 |

**7.** **a)** What is Caroline's monthly take-home pay?
**b)** What do her monthly expenses total?

**8.** Assume Caroline has no starting balance from the previous month. Does she have enough money from her first pay to cover all expenses due before her second pay?

**9.** What percent of Caroline's monthly take-home pay are her monthly expenses?

**10.** What percent of her total monthly expenses is due before her second pay?

**11.** Assume Caroline is paid $1992 once a month. How would this timing affect her bill payments if her pay day was

**a)** October 1? **b)** October 15? **c)** October 31?

**12.** Lorne's due dates for bills are spaced throughout the month. He receives one pay cheque monthly, on the first Friday of the month. He is pleased because he has lots of money to spend on non-essentials. What advice would you give him?

**13.** How often would you prefer to be paid? Why?

# 2.7 Purchasing Power

Josephine is thinking about changing jobs. She wants to increase her **purchasing power**, or financial ability to make purchases.

She would like to
- move to an apartment of her own
- own and operate a used car
- have more money for non-essential expenses, such as entertainment, a vacation, and charitable donations

## Develop

**1.** Josephine now works as a sales clerk at a hardware store. It is close to the apartment that she shares with a friend. She has

an hourly pay rate of $7.25
six 6-hour shifts a week, one on the weekend
no benefits

| deductions for income tax | $23.50 |
|---|---|
| EI | $5.87 |
| CPP | $8.33 |

**a)** What is Josephine's gross weekly pay?
**b)** What is her net weekly pay?
**c)** What is her net monthly pay? Hint: Multiply her net weekly pay by 52. Then divide by 12.

**2.** Josephine's current monthly living expenses are as follows:

| | |
|---|---|
| rent, including utilities | $420 |
| phone | $19 |
| cable | $22 |
| groceries | $150 |
| credit card payment | $250 |
| medical | $28 |

Her credit card payment covers purchases of clothes, gifts, and household items. Her medical expense is the monthly amount budgeted for her dental checkup, prescriptions, and glasses.

**a)** What is the total of Josephine's monthly living expenses?
**b)** How much money does Josephine have left at the end of a month?

## Practise

3. Josephine could rent an apartment for herself for $675 a month, including utilities.
   a) Which of her monthly expenses are likely to double if she lives alone? Explain.
   b) How much would living in this apartment actually cost her a month?

4. Josephine could buy a car with monthly expenses of

| | |
|---|---|
| payments | $235 |
| insurance | $150 |
| fuel and maintenance | $60 |

   a) How much would owning and operating this car cost her each month?
   b) Check her monthly expenses. Would owning a car save her any money? Explain.

5. How much more does Josephine's net monthly pay need to be for her to afford an apartment on her own as well as a car?

6. Without even looking into other jobs, Josephine decides that if she really wants a car, which she does, she had better stay in her shared apartment. She also comes to realize that some bigger non-essential expenses, like a vacation, will have to wait. How do you think Josephine came to this conclusion? Do you agree?

7. To own and operate a car, Josephine decides that her net monthly pay needs to be at least $450 higher. How did she decide on $450?

8. Josephine starts searching for jobs and finds three different sales positions, all with salaries plus commission. The typical sales are based on the average sales of other staff for a month.
   a) Explain how the amounts given in cells E2 and F2 were calculated.
   AT b) FILL DOWN Columns E and F, Rows 2 to 5. What are the gross monthly incomes from the other stores?

| | A | B | C | D | E | F |
|---|---|---|---|---|---|---|
| 1 | Store | Salary ($) | Rate of Commission (%) | Typical Sales ($) | Commission ($) | Gross Monthly Income ($) |
| 2 | Guru | 1000 | 4 | 4100 | 164.00 | 1164.00 |
| 3 | Amy's | 700 | 5 | 5040 | | |
| 4 | Marci's | 1300 | 2 | 4000 | | |
| 5 | Expressions | 800 | 3 | 4750 | | |

9. None of these sales positions has a health plan, something that Josephine wants. Why would a health plan appeal to her?

**10.** Josephine does not know the exact deductions for these sales positions. She estimates that they would be 15% of gross income.

**a)** Explain how the amounts given in cells G2 and H2 were calculated.

AT **b)** FILL DOWN Columns G and H, Rows 2 to 5. What are the net monthly incomes from the other stores?

| | A | B | C | D | E | F | G | H |
|---|---|---|---|---|---|---|---|---|
| 1 | Store | Salary ($) | Rate of Commission (%) | Typical Sales ($) | Commission ($) | Gross Monthly Income ($) | Deductions ($) | Net Monthly Income ($) |
| 2 | Guru | 1000 | 4 | 4100 | 164.00 | 1164.00 | 174.60 | 989.40 |
| 3 | Amy's | 700 | 5 | 5040 | | | | |
| 4 | Marci's | 1300 | 2 | 4000 | | | | |
| 5 | Expressions | 800 | 3 | 4750 | | | | |

**11.** Which, if any, of the sales jobs would help Josephine increase her purchasing power? Why?

WS **12.** Josephine finds three other positions that interest her and seem to suit her qualifications. Find the gross monthly income for each position.

| | Position | Method of Payment | Weekly Hours Worked | Benefits |
|---|---|---|---|---|
| **a)** | office assistant | annual salary of $19 200 | 37.5 | health plan |
| **b)** | telemarketer | hourly rate of $10.95 | 32 | none |
| **c)** | custodian | hourly rate of $12.35 | 35 | health plan |

WS **13.** Josephine does not know the exact deductions that apply to the positions in question 12. She estimates that they would be 15% of gross income for the first two positions. Using that estimate, determine the deductions and net monthly income for each of them.

WS **14.** The custodial job is unionized. Josephine estimates that deductions, which include union dues, might be 17% of gross income. Using that estimate, determine the deductions and net monthly income.

**15.** Which, if any, of these positions would help Josephine increase her purchasing power? Why?

**16.** Consider all seven jobs.

**a)** What jobs do you recommend that Josephine apply for? Why?

**b)** Assume that Josephine applies, gets interviews, and is offered the jobs. What other information should she have found out about the jobs to help her decide which to accept?

# 2.8 Putting It All Together: Deductions and Living Expenses

**1.** Refer to your work from Section 1.8, where you applied for a job and got it.
  **a)** What is your job?
  **b)** What did you expect to earn in a week, a month, and a year at that job?
  **c)** Is that gross income or net income?
  **d)** Complete a Personal Tax Credits Return and determine your claim code.

**2.** Use tables from your teacher to determine your weekly deductions for income taxes, EI, and CPP.

**3.** Would any other deductions, such as union dues, affect your pay cheque? If so, what might they be? Research or estimate how much they would be.

**4. a)** What do your weekly deductions total?
  **b)** What percent of your gross weekly income are the deductions?
  **c)** Calculate your take-home pay for a week and for a month.

**5.** Review your work from Section 2.4 and create a list of living expenses for yourself.
- Assume that you are renting, either on your own or sharing.
- Classify the expenses as essential or non-essential.
- Include the monthly amount of the expenses that you found.

**6.** How frequently would you prefer to be paid to help you with meeting your expenses throughout the month? Why?

**7. a)** Does your job give you an adequate pay cheque to meet all your expenses?
  **b)** If you could not meet all your expenses, what expenses could you reduce or eliminate? Explain.

**8.** Why would you want to save some money each month? Adjust your expenses as needed to save at least $75 a month. What percent of your gross monthly income will your savings be?

# 2.9 Chapter Review

1. Why does a new employee complete a TD1 form, a Personal Tax Credits Return?

2. Why is income tax deducted by the federal and provincial governments from our gross pay?

3. **a)** Which is often called take-home pay—gross or net pay? Explain.
   **b)** Name four possible payroll deductions beyond the standard government deductions.
   **c)** What is it called when an employer pays either in part or in full for such items as a health plan?

4. Calculate the net weekly pay for each position.

**a)**

| | | |
|---|---|---|
| gross pay | $467.98 | |
| deductions | federal income tax | $47.50 |
| | provincial income tax | $18.40 |
| | EI | $10.53 |
| | CPP | $17.22 |
| | union dues | $25.00 |
| | health plan | $18.50 |

**b)**

| | | |
|---|---|---|
| gross pay | $454.50 | |
| deductions | federal income tax | $45.70 |
| | provincial income tax | $17.70 |
| | EI | $10.23 |
| | CPP | $16.67 |
| | union dues | $22.00 |

**c)**

| | | |
|---|---|---|
| gross pay | $10.58 per hour for 40 hours | |
| deductions | federal income tax | $40.30 |
| | provincial income tax | $15.85 |
| | EI | $9.52 |
| | CPP | $15.30 |

5. What is usually the highest monthly expense? Why does it vary greatly from person to person?

6. Name four essential and four non-essential monthly expenses. Justify your choices.

7. Describe two employees' situations—their earnings and expenses. Which person has greater purchasing power? Justify your choice.

**8.** Marco's semimonthly take-home pay is $884. He is paid November 1 and November 15. His November expenses are as follows:

| | | |
|---|---|---|
| phone | $35 | November 2 |
| groceries | $50 | November 5 |
| credit card payment | $389 | November 11 |
| groceries | $50 | November 12 |
| car payment | $227 | November 14 |
| groceries | $50 | November 19 |
| cable | $42 | November 20 |
| car insurance | $120 | November 25 |
| groceries | $50 | November 27 |
| rent, including utilities | $745 | November 30 |

**a)** How much is his monthly take-home pay?
**b)** What do his monthly expenses total?
**c)** Assume Marco has no starting balance. Does he have enough money from his November 1 pay to cover all the expenses due before his next pay?
**d)** What percent of Marco's monthly take-home pay are his monthly expenses?

**9.** What are three different pay frequencies? Which do you think would allow you to manage the payment of expenses most easily? Explain.

**10.** Jeff has a job as a furniture salesperson. He has

a monthly salary of $1050
3% commission on sales, with an average of $10 000 in sales a month
no benefits

| | |
|---|---|
| deductions for income tax | $157.20 |
| EI | $30.36 |
| CPP | $46.48 |

His living expenses are

| | |
|---|---|
| rent, including utilities | $550 |
| phone | $32 |
| groceries | $180 |
| credit card payment | $250 |
| bus pass | $38 |

His credit card payment covers purchases of clothes, gifts, and household items.

Jeff would like to buy a used car. Assess his situation and explain why he can or cannot afford to own and operate a used car.

CHAPTER

# 3 Paying Taxes

In this chapter, you will

- identify the information needed for filing an income tax return and discuss why it is needed
- investigate how to get help completing an income tax return
- solve problems estimating and calculating Provincial Sales Tax (PST) and Goods and Services Tax (GST)
- investigate other forms of taxation

Near the end of the chapter, you will

- determine the annual income tax, CPP, and EI deductions that you would pay based on working one year at the job you found in Chapter 1
- select items you would like to purchase and determine their cost, including taxes

# 3.1 Information for Filing Income Taxes

In Chapter 2 you learned that income tax is deducted from each pay cheque and that the amount deducted depends upon the information entered on the TD1 form.

## Explore

What does it mean to file a personal income tax return? Who is required to file a personal income tax return and by when?

## Develop

**1.** To file a personal income tax return, the following are needed:
- T1 General Forms
- General Income Tax and Benefit Guide
- information slips
- receipts for deductions or credits to be claimed

The **T1 General Forms** and **General Income Tax and Benefit Guide** are mailed to people who filed income tax returns the previous year. Otherwise, they can be obtained at postal outlets or on-line. The forms are blank. Tax filers complete applicable parts using information in the guide.

**a)** Why would the government provide forms for tax filers to complete when filing income tax returns?

**b)** Why would those forms come with a guide?

2. **Information slips** are also government forms. Unlike T1 General Forms, they are already completed when taxpayers receive them. They include T4 and T5 slips.

   A **T4** information slip is completed and issued by employers. It is a **Statement of Remuneration Paid**. That means it is an official record of
   - the amount of income earned during the year
   - the amount of income tax, Canada Pension Plan, and Employment Insurance paid through payroll deductions
   - other amounts paid through payroll deductions that affect income tax

   People who held more than one job are issued a T4 slip by each employer.

   Why do you think employers are required to complete and issue T4 information slips?

3. A **T5** information slip is a **Statement of Investment Income**. It is completed and issued by banks and other financial institutions to people who invested or saved money that earned interest. Interest is income and is subject to income tax. People who earned interest at more than one financial institution receive a T5 slip from each one.

   Why do you think financial institutions are required to complete and issue T5 information slips?

4. Receipts that can be used for tax deductions or credits include those for
   - charitable donations
   - medical expenses
   - union dues
   - child-care expenses
   - moving expenses
   - Registered Retirement Savings Plan (RRSP) contributions

   When such expenses meet specific conditions, they reduce the amount of income tax paid.

   **a)** Where would you get receipts for expenses such as moving and medical?

   **b)** Why should you keep receipts for such expenses?

## Practise

5. Georgina works part time at Tasty Hamburger Joint. Here is the T4 slip that Georgina's employer sent her.

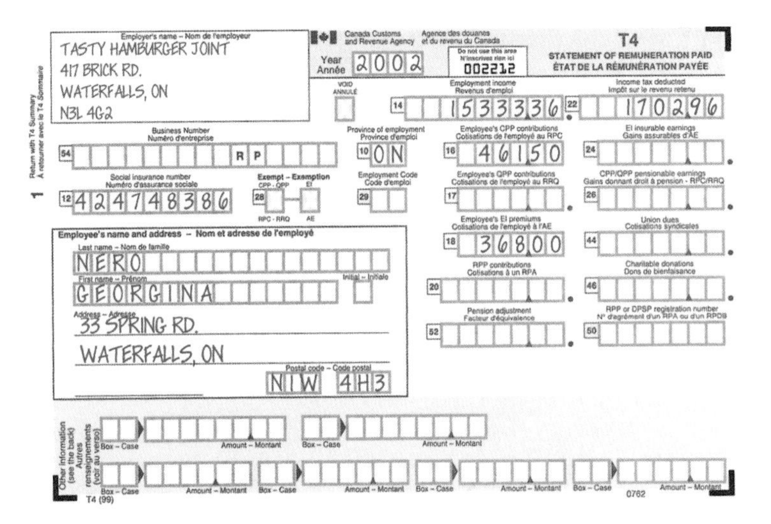

Employer's name – Nom de l'employeur
TASTY HAMBURGER JOINT
417 BRICK RD.
WATERFALLS, ON
N3L 4G2

Canada Customs and Revenue Agency / Agence des douanes et du revenu du Canada

Year / Année: 2002

Do not use this area / N'inscrivez rien ici: 002212

T4
STATEMENT OF REMUNERATION PAID
ÉTAT DE LA RÉMUNÉRATION PAYÉE

Return with T4 Summary / À retourner avec le T4 Sommaire

|  |  |  |
|---|---|---|
| | VOID / ANNULÉ | |
| 14 | Employment income / Revenus d'emploi | 1533336 |
| 22 | Income tax deducted / Impôt sur le revenu retenu | 170296 |
| 54 | Business Number / Numéro d'entreprise | RP |
| 10 | Province of employment / Province d'emploi | ON |
| 16 | Employee's CPP contributions / Cotisations de l'employé au RPC | 46150 |
| 24 | EI insurable earnings / Gains assurables d'AE | |
| 12 | Social insurance number / Numéro d'assurance sociale | 424748386 |
| 28 | Exempt – Exemption: CPP-QPP / RPC-RRQ; EI / AE | |
| 29 | Employment Code / Code d'emploi | |
| 17 | Employee's QPP contributions / Cotisations de l'employé au RRQ | |
| 26 | CPP/QPP pensionable earnings / Gains donnant droit à pension - RPC/RRQ | |
| 18 | Employee's EI premiums / Cotisations de l'employé à l'AE | 36800 |
| 44 | Union dues / Cotisations syndicales | |
| 20 | RPP contributions / Cotisations à un RPA | |
| 46 | Charitable donations / Dons de bienfaisance | |
| 52 | Pension adjustment / Facteur d'équivalence | |
| 50 | RPP or DPSP registration number / N° d'agrément d'un RPA ou d'un RPDB | |

Employee's name and address – Nom et adresse de l'employé
Last name – Nom de famille: NERO
First name – Prénom: GEORGINA
Initial – Initiale:
Address – Adresse: 33 SPRING RD.
WATERFALLS, ON
Postal code – Code postal: N1W 4H3

Other information (see the back) / Autres renseignements (voir au verso)
Box – Case | Amount – Montant

T4 (99) 0762

a) What was Georgina's income for the year?
b) How much did she have deducted for each of the following?
   i) income tax
   ii) Canada Pension Plan
   iii) Employment Insurance
   iv) union dues
   v) charitable donations
c) Georgina donated to charities, but not through payroll deductions. Where would she get those receipts?

6. Where does a taxpayer get each of the following items?
   a) T1 General Forms and General Income Tax and Benefit Guide
   b) T4 information slip
   c) T5 information slip
   d) receipts to be used for tax deductions or credits, such as child care

7. Jeremy received two T4 information slips. What does that indicate?

**8. a)** Have you received a T4 information slip? From where did you get it?

**b)** Have you filed an income tax return? For what year did you file it?

**9.** Shamir has money in a savings account. Here is the T5 slip that Shamir's bank sent to him.

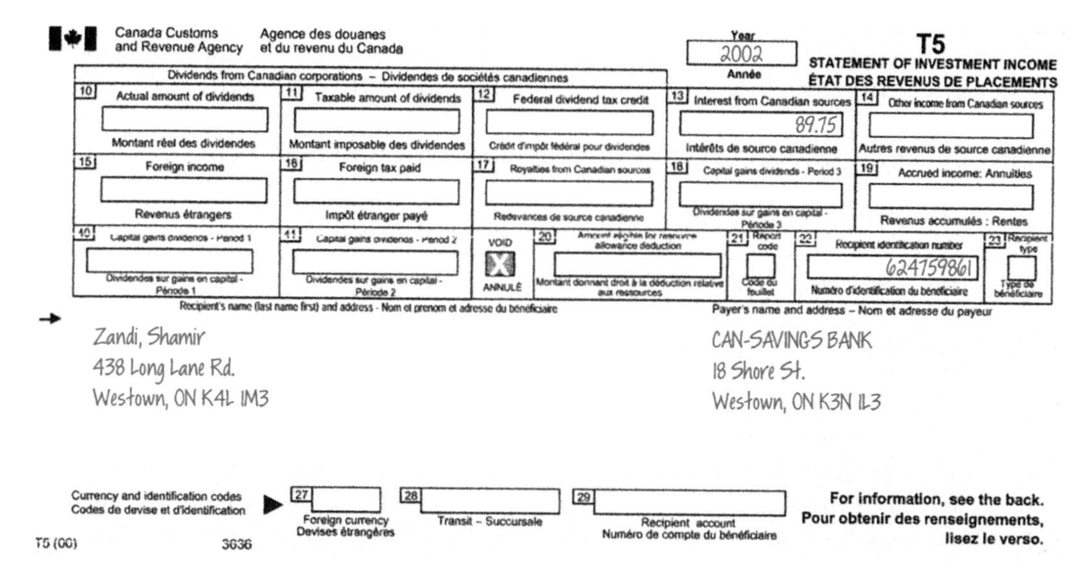

Canada Customs and Revenue Agency / Agence des douanes et du revenu du Canada

Year / Année: 2002

**T5**
**STATEMENT OF INVESTMENT INCOME**
**ÉTAT DES REVENUS DE PLACEMENTS**

Dividends from Canadian corporations – Dividendes de sociétés canadiennes

|  |  |  |
|---|---|---|
| 10 | Actual amount of dividends / Montant réel des dividendes | |
| 11 | Taxable amount of dividends / Montant imposable des dividendes | |
| 12 | Federal dividend tax credit / Crédit d'impôt fédéral pour dividendes | |
| 13 | Interest from Canadian sources / Intérêts de source canadienne | 89.75 |
| 14 | Other income from Canadian sources / Autres revenus de source canadienne | |
| 15 | Foreign income / Revenus étrangers | |
| 16 | Foreign tax paid / Impôt étranger payé | |
| 17 | Royalties from Canadian sources / Redevances de source canadienne | |
| 18 | Capital gains dividends - Period 3 / Dividendes sur gains en capital - Période 3 | |
| 19 | Accrued income: Annuities / Revenus accumulés : Rentes | |
| 40 | Capital gains dividends - Period 1 / Dividendes sur gains en capital - Période 1 | |
| 41 | Capital gains dividends - Period 2 / Dividendes sur gains en capital - Période 2 | |
| | VOID / ANNULÉ | X |
| 20 | Amount eligible for resource allowance deduction / Montant donnant droit à la déduction relative aux ressources | |
| 21 | Report code / Code du feuillet | |
| 22 | Recipient identification number / Numéro d'identification du bénéficiaire | 624759861 |
| 23 | Recipient type / Type de bénéficiaire | |

Recipient's name (last name first) and address - Nom et prénom et adresse du bénéficiaire

Zandi, Shamir
438 Long Lane Rd.
Westown, ON K4L 1M3

Payer's name and address – Nom et adresse du payeur

CAN-SAVINGS BANK
18 Shore St.
Westown, ON K3N 1L3

Currency and identification codes / Codes de devise et d'identification

| 27 | 28 | 29 |
|---|---|---|
| Foreign currency / Devises étrangères | Transit – Succursale | Recipient account / Numéro de compte du bénéficiaire |

**For information, see the back.**
**Pour obtenir des renseignements, lisez le verso.**

T5 (00) 3636

How much interest did Shamir earn at Can-Savings Bank?

**10.** Georgina's Social Insurance Number (SIN) appeared on her T4 slip and Shamir's SIN appeared (as Recipient indentification number) on his T5 slip. Discuss why.

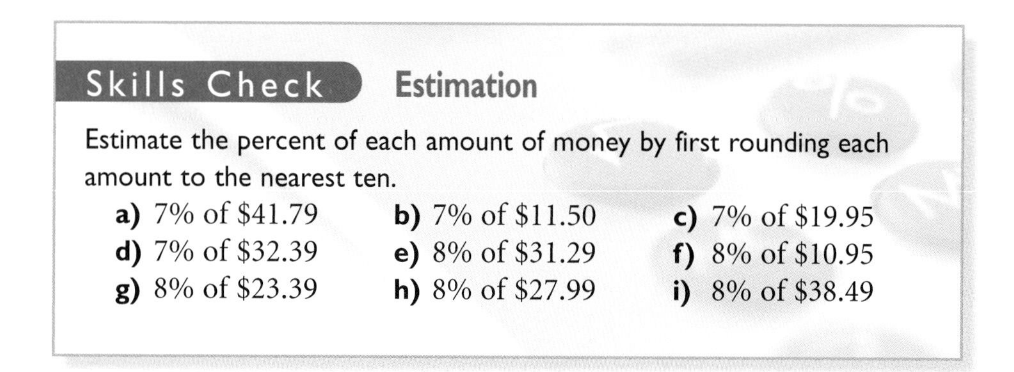

## Skills Check **Estimation**

Estimate the percent of each amount of money by first rounding each amount to the nearest ten.

**a)** 7% of \$41.79 **b)** 7% of \$11.50 **c)** 7% of \$19.95

**d)** 7% of \$32.39 **e)** 8% of \$31.29 **f)** 8% of \$10.95

**g)** 8% of \$23.39 **h)** 8% of \$27.99 **i)** 8% of \$38.49

# 3.2 Help with Income Taxes

The first time you file a return or deal with changes to sources of income or deductions, you may need more help than what is provided in the General Income Tax and Benefit Guide.

## Develop

**1.** The government department that deals with income tax is the **Canada Customs and Revenue Agency (CCRA)**. Inside the General Income Tax and Benefit Guide, you will find additional ways to get information on how to file a tax return:

- over the Internet, using the CCRA Web site
- over the telephone, using their toll-free numbers
- in person, by visiting a tax services office (You can find the nearest tax services office by checking the government section of your telephone book or their Web site.)

The Guide also indicates that the government trains volunteers to complete basic tax returns for low-income individuals. A phone number is provided to get more information about this free service.

What would be some advantages and disadvantages of each additional way to get information on how to file a tax return?

**2.** There are other sources of help for filing tax returns.

- Libraries or community centres often offer classes in January or February to help people complete their income tax forms.
- A business whose sole purpose is to complete tax returns can be hired.
- An accounting firm can be hired to complete income tax returns.
- Tax return software can be purchased.

Select one of the sources of help and research the following about it.

**a)** availability in your community
**b)** names, addresses, and phone numbers
**c)** cost of getting a basic tax return completed
**d)** ways that returns are filed—electronically or by regular mail
**e)** advantages and disadvantages of the source

**3.** Present your findings from question 2 to the class and identify where you obtained your information.

**4.** Which source of help for filing a return, if any, would you use? Why?

# 3.3 Provincial and Federal Sales Taxes

Another form of taxation that governments use to collect money to pay for services that they provide is sales taxes.

**Provincial Sales Tax (PST)** is collected on most goods and some services. In Ontario PST is 8% of the selling price. Some PST exempt items include

- groceries
- children's clothing
- books and newspapers
- hair styling
- prepared food costing $4 or less
- footwear costing $30 or less
- dry cleaning
- car washing

The federal tax, **Goods and Services Tax (GST)**, is collected on most goods and services. GST is 7% of the selling price. Some GST exempt items include

- groceries
- child care
- bridge, road, and ferry tolls

## Explore

You buy a pair of running shoes with a selling price of $69.95. The total cost you pay is $80.45. How could the total cost have been calculated?

## Develop

### Example

Estimate and then calculate the taxes and the total cost of a CD with a selling price of $19.75.

#### Solution

**Estimate.**

GST
7% of $19.75 ≐ 1.40

Round, multiply by 7, and divide by 100.
7 × 20 = 140 and 140 ÷ 100 = 1.40

PST
8% of $19.75 ≐ 1.60

Round, multiply by 8, and divide by 100.
8 × 20 = 160 and 160 ÷ 100 = 1.60

Total cost
$20 + 1.40 + 1.60 = 20 + 3$
$= 23$

The total cost is about $23.

**Calculate.** 

GST

$7\%$ of $\$19.75 = 0.07 \times 19.75$
$\doteq 1.38$

PST

$8\%$ of $\$19.75 = 0.08 \times 19.75$
$= 1.58$

Total cost

$19.75 + 1.38 + 1.58 = 22.71$

The total cost is exactly $22.71.

## Practise

**1.** Estimate the GST on each purchase.

**a)** $62 **b)** $498 **c)** $245
**d)** $10.99 **e)** $20.49 **f)** $12.39

**2.** Estimate the PST on each purchase.

**a)** $79 **b)** $398 **c)** $479
**d)** $24.99 **e)** $119.75 **f)** $88.50

**3.** Estimate the taxes and total cost of a pair of shoes that sells for $29.49 and is subject to PST only.

**4.** Each of the following labour charges is subject to GST only. Estimate and calculate the total cost of each labour charge.

**a)** labour for installing flooring, $198.50
**b)** labour for central air conditioner repair, $72
**c)** labour for plumbing, $138
**d)** labour for furnace maintenance, $56

**5.** For each purchase shown below,

**i)** estimate the GST
**ii)** estimate the PST
**iii)** estimate the total cost
**iv)** calculate the GST
**v)** calculate the PST
**vi)** calculate the total cost

**a)**

**b)**

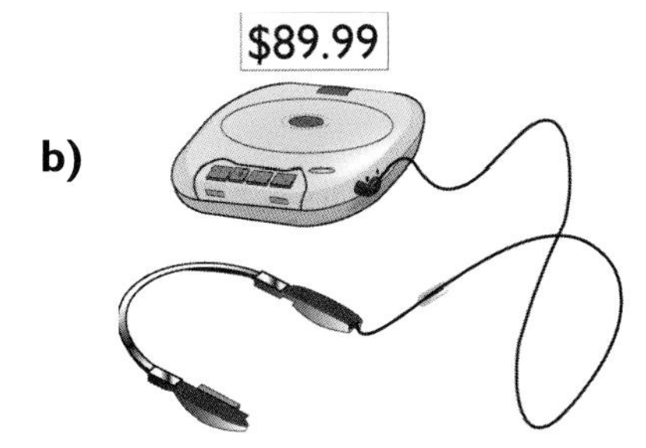


**c)**

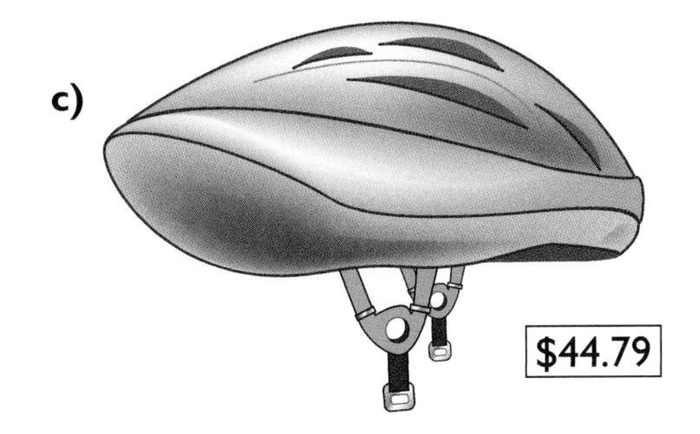


**d)**

**e)**

**f)**

**6.** What percent of the selling price is both taxes together? Explain.

**7.** The total cost of a telephone subject to both taxes is $103.44. Is the price of the telephone before taxes $89.95 or $87.92? Show your work.

### Skills Check — One Number as a Percent of Another

Calculate the first amount as a percent of the second. Round each percent to the nearest whole number.

**a)** 15¢ 70¢ **b)** 16¢ 77¢ **c)** 12¢ 65¢
**d)** 19¢ 50¢ **e)** 15.5¢ 67.9¢ **f)** 18.5¢ 72.9¢

# 3.4 Other Forms of Taxation

We pay both direct taxes and indirect taxes. The taxes discussed in the previous sections—income taxes, PST, and GST—are **direct taxes** because the exact amount of tax collected is evident.

## Explore

What other taxes are you aware of? Are they direct taxes?

## Develop

**1.** Another direct tax is **property tax**. Owners of real estate, land with or without buildings, pay annual taxes. Property tax is a percent of the value of the property.

Taxes are collected by the federal, provincial, and municipal governments. How could you find out which level of government collects property taxes?

**2.** **Indirect taxes** are collected on such goods as gasoline, cigarettes, alcohol, and jewellery. These taxes are included in the selling price, not added to it.

An indirect **jewellery tax** is collected by the federal government. It is 10% of the wholesale price, or price that a store pays. It is included in the store's selling price.

Without using a calculator, determine the 10% tax on the following jewellery.

**a)** a watch, with a wholesale price of $78.50
**b)** a ring, with a wholesale price of $305
**c)** a necklace, with a wholesale price of $250

**3.** In addition to the PST and GST added to the selling price of cigarettes, three **cigarette taxes** are included in the selling price. On a carton of 200 cigarettes with a selling price of $40.70,

- the provincial government collects $8.90
- the federal government collects $6.85
- the federal government collects another $5.50

**a)** What is the total of the three cigarette taxes on one carton?
**b)** What is the price excluding the three cigarette taxes?
**c)** What percent of the selling price is the total of the three cigarette taxes?

**Round all amounts of money to the nearest cent.**
**Round all percents to the nearest whole number.**

4. Assume the selling price of a carton of 200 cigarettes is $40.70.
   **a)** How much is the PST on a carton?
   **b)** How much is the GST on a carton?
   **c)** What is the total cost of a carton of cigarettes, including PST and GST?
   **d)** How much of the total cost of a carton is made up of the five taxes—PST, GST, and the three taxes from question 3?
   **e)** How much of the total cost of one cigarette is made up of the five taxes?

5. Why do you think cigarettes are taxed so highly?

6. No PST is added to the selling price of gasoline. Although GST is usually added to the selling price, it is included in the selling price for gasoline. As well, **gasoline taxes** are collected by both the federal and provincial governments. The federal government collects 10¢/L and the provincial government collects 14.7¢/L.

   What is the total of the two gasoline taxes on each amount of gasoline?

   **a)** 1 L **b)** 40 L **c)** 30 L **d)** 44.5 L

7. What percent of each selling price is the 10¢/L tax collected by the federal government?

   **a)** 65.7¢/L **b)** 73.9¢/L

8. What percent of each selling price is the 14.7¢/L tax collected by the provincial government?

   **a)** 65.7¢/L **b)** 73.9¢/L

9. Use the current selling prices of gasoline in your area. What percent of those selling prices is each of the following?
   **a)** the 10¢/L tax collected by the federal government
   **b)** the 14.7¢/L tax collected by the provincial government
   **c)** the combined gasoline taxes collected by both governments

**Skills Check** **Mental Math**

Calculate each cost mentally.

**a)** 8 yogurt at 59¢ each **b)** 5 apples at 29¢ each
**c)** 7 m of ribbon at 41¢/m **d)** 6 m of ribbon at 88¢/m
**e)** 4 oranges at 52¢ each **f)** 7 pears at 38¢ each

# 3.5 Career Focus: Gas Bar Attendant

Nancy is a gas bar attendant. The duties of her job include

- greeting customers
- asking them what quantity and grade of gasoline they would like
- inquiring whether they would like their oil checked
- pumping gas
- checking oil and adding oil, if necessary
- cleaning windshields
- using a cash register to calculate customers' bills
- accepting payments in cash, by debit card, and by credit card
- checking gasoline levels in underground tanks
- inputting data into a computer and reading reports

**Round all amounts of money to the nearest cent.**

**1.** What characteristics do you think are important for the part of Nancy's job that calls for contact with customers? Explain your choices.

**2.** One customer wants a fill-up with regular unleaded gas and the oil checked. His car needs a litre of oil. As well, he buys a can of pop. The fill-up comes to \$32, the oil is \$3.40, and the pop is \$1.25. The oil and pop are subject to PST and GST.

**a)** What is the customer's total bill?
**b)** How much change must be given if \$40 is offered?

**3.** The quantity of regular unleaded gas in question 2 was 43.9 L. It sells for 72.9¢/L. How much of the \$32 for the fill-up was each of the following taxes?

**a)** the federal 10¢/L gasoline tax
**b)** the provincial 14.7¢/L gasoline tax

4. At the end of her shift, Nancy checks the amount of gasoline in the underground tanks. She uses a 400 cm dipstick to measure the depth of gasoline remaining in the tanks. She inputs the measurement in a computer and gets a report back. The report indicates how many litres of gasoline remain in the tank.

| **Tank 01** | |
|---|---|
| **CA6 Regular Unleaded** | |
| Temperature (°C) | 15 |
| Centimetres | 152 |
| Litres | 25 188 |

   a) Which grade of gasoline was in Tank 01?
   b) Which number in the report is the measurement that Nancy input?
   c) When the tank is full, the depth of gasoline is 350 cm. How many centimetres of gasoline were pumped out since the tank was last filled?

5. When a depth in centimetres is input, the computer calculates the volume in litres. In Tank 01 in question 4, the computer calculates 25 188 L when a depth of 152 cm is input.
   a) How many litres would the computer calculate for a depth of 1 cm? Round your answer to one decimal place.
   b) What assumption about the shape of the tank did you make to answer part a)?
   c) Determine how many litres of gasoline were pumped out since the tank was last filled by calculating the number of litres that the computer would calculate for the number of centimetres you found in question 4 c).
   d) Assume that the gasoline pumped out of Tank 01 was sold for 72.9¢/L. What amount of payment was received from the sale of the gas sold since the tank was last filled?
   e) How much federal gasoline tax at 10¢/L was paid on the gas sold since the tank was last filled?
   f) How much provincial gasoline tax at 14.7¢/L was paid on the gas sold since the tank was last filled?

# 3.6 Putting It All Together: Paying Taxes

As you have seen, we pay direct and indirect taxes.

**Recall "your job" from Chapters 1 and 2.**
**Round all amounts of money to the nearest cent.**

**1.** Give three examples of each type of taxes.
  **a)** direct
  **b)** indirect

**2.** Why do we pay taxes?

**3.** Refer to your work from questions 1 and 2 in Section 2.8. In that question 2, you determined weekly payroll deductions that applied to "your job." These were
- federal income tax
- provincial income tax
- CPP
- EI

  **a)** How much money did you determine that you would earn a year?
  **b)** Calculate the total amount of each deduction that would appear on your T4 slip.

WS **4.** Use newspaper ads, flyers, and catalogues.
  **a)** Select four items that you would like to purchase for yourself, your family, or someone else. Select from at least two of these categories: clothing, furniture, electronics, entertainment tickets.
  **b)** Estimate the PST on each item.
  **c)** Estimate the GST on each item.
  **d)** Estimate the after-tax cost of each item.
  **e)** Calculate the PST on each item.
  **f)** Calculate the GST on each item.
  **g)** Calculate the after-tax cost of each item.

**5.** Which of the items that you selected in question 4 do you think you could afford? Base your decision on the earnings and deductions from "your job." Justify your choices.

# 3.7 Chapter Review

**1.** T1 General Forms and the General Income Tax and Benefit Guide are needed to file an income tax return.

**a)** What other documents are needed?
**b)** Where do you obtain the documents you named in part a)?
**c)** Why do you need those documents?

**2. a)** Name two sources of help for filing income taxes.
**b)** Give an advantage and disadvantage of each.

**3. a)** What are four taxes other than income taxes that are collected?
**b)** Name four services that are paid for by taxes.

**4. a)** What do the initials PST and GST stand for?
**b)** Identify three differences between PST and GST.

WS **5.** Each purchase is subject to both PST and GST. Complete the table. When estimating, round to numbers that are easy to multiply.

| | Item | Selling Price | Estimate of PST | Estimate of GST | Estimate of Total Cost | Actual PST | Actual GST | Actual Total Cost |
|---|---|---|---|---|---|---|---|---|
| a) | roller blades | $125.50 | | | | | | |
| b) | headphones | $49.99 | | | | | | |
| c) | shirt | $38.98 | | | | | | |
| d) | camera | $189.79 | | | | | | |
| e) | cross-country skis | $229.89 | | | | | | |

**6. a)** What is the difference between direct and indirect taxes?
**b)** Give two examples of each type of taxes.

**7. a)** How much is the 10% jewellery tax on a ring with a wholesale price of $411?
**b)** Gasoline taxes are 10¢/L and 14.7¢/L. The selling price is 74.6¢/L. What percent of the selling price is the total of the two gasoline taxes?

**8.** Cigarettes are subject to GST and PST. To how many indirect taxes are they subject?

**9.** The labour portion of Joan's furnace repair bill is $58.10. The cost of parts is $27.53. This labour is subject to GST only. The parts are subject to PST and GST. Determine if $100 is enough to pay Joan's bill.

CHAPTER

# 4 Making Purchases

In this chapter, you will examine aspects of making cash purchases, such as

- providing the correct change for amounts offered
- deciding amounts to offer for purchases to get back fewer coins in your change
- estimating and calculating PST, GST, and total costs
- estimating and calculating discounts and sale prices
- estimating and calculating total costs, including taxes, on discounted purchases

Near the end of the chapter, you will

- select items to purchase
- estimate and calculate discounts where applicable
- estimate and calculate taxes where applicable
- estimate and calculate total costs

# 4.1 Making Change

If you work in a store, the cash register will calculate how much change you should give customers. However, it will not tell you what bills and coins to give.

## Explore

For a purchase totalling $1.14, what coins would you give as change to a customer who offers you a $2 coin? What if the customer offered you a $2 coin and a quarter?

## Develop

### Example

What change would you give each customer?

**a)** For an $8.17 purchase, one customer tenders (or offers) a $10 bill. You enter 10 and the cash register calculates the change.

$10 - 8.17 = 1.83$ amount tendered − purchase

The customer is owed $1.83.

**b)** Another customer who makes the same $8.17 purchase tenders a $10 bill and two dimes.
You enter 10.20 and the cash register calculates the change.

$10.20 - 8.17 = 2.03$ amount tendered − purchase

The customer is owed $2.03.

```
***Your Town Food Store***
1234 Main Street
Townsville, ON

yogurt          2.69
blueberries     3.29
sugar           2.19
TOTAL           8.17
CASH           10.00
CHANGE DUE      1.83

G.S.T.# 123456789RT
10/15/02  11:46
PLEASE COME AGAIN!
```

### Solution

**a)** If possible, you give the customer the fewest coins in change.
Since the change is less than $2, the largest coin is a $1 coin.
That leaves 83¢.
The next largest coin is a quarter, and 3 quarters is 75¢.
That leaves 8¢.
The next largest coin is a nickel.
That leaves 3¢—3 pennies.
You can check by counting out the change.
$1—1.25—1.50—1.75—1.80—1.81—1.82—1.83

**b)** If possible, you give the customer the fewest coins in change.
Since the change is just over $2, the largest bill or coin is a $2 coin.
That leaves 3¢—3 pennies.
You can check by counting out the change.
$2—2.01—2.02—2.03

## Practise

1. A customer tendered a $5 bill, a quarter, and a nickel for a card with a total cost of $3.28.
   a) What amount tendered should be entered in the cash register?
   b) What change does the cash register calculate?
   c) What bills and coins should this customer receive?
   d) Check your answer to part c) by counting out the change.
   e) Why do you think the customer gave the cashier additional coins?

**For the situations in questions 2 to 11, answer the following.**
**a) What amount tendered should be entered in the cash register?**
**b) What change does the cash register calculate?**
**c) What bills and coins should the customer receive?**
**d) Check your answer to part c) by counting out the change.**

2. Mayda offers a $10 bill and 2 quarters to pay her lunch bill of $8.37.
3. Serita tenders two $20 bills to pay $26.95 for a bicycle pump.
4. Joseph offers a $20 bill and a $10 bill for his purchase of $21.92 worth of groceries.
5. Juanita tenders a $10 bill and a $5 bill to pay for an $11.50 purchase.
6. Rod tenders a $10 bill, a $5 bill, a dime, and 2 pennies to pay for stationery supplies costing $13.87.

| | Total Bill | Amount Tendered |
|---|---|---|
| 7. | $8.49 |  |
| 8. | $9.14 |  |
| 9. | $12.37 |  |
| 10. | $15.61 |  |
| 11. | $18.44 |  |

12. The total bill being paid is $8.63. If the object is to give the customer the fewest coins, which coins should you use for each amount tendered?
    a) $10 b) $9 c) $8.75 d) $8.70 e) $8.65

# 4.2 Getting Back Fewer Coins

In Section 4.1, you saw that some people give combinations of bills and coins to reduce the number of coins they receive in change. When they do that, they have fewer coins to carry and the cashier has more coins for making change for other customers.

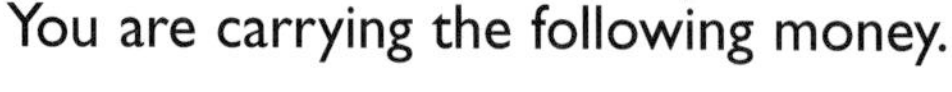

## Explore

You are carrying the following money.

You make a purchase totalling $3.53. What would you offer to get back the fewest coins?

## Develop

For a bill of $11.29, consider the following.

- The exact amount of $11.29 would result in no change.
- Amounts of $11.30, $11.35, and $11.50 would result in small coin change.
- Amounts of $12.29, $15.29, and $20.29 would result in full dollar amounts of change.
- Amounts of $12.30, $15.30, and $20.30 would result in full dollar amounts and one penny in change.
- Amounts of $12.35, $12.50, $15.35, $15.50, $20.35, and $20.50 would result in full dollar amounts and small coin change.

What you offer depends on what coins and bills you have.

### Example

Rajat orders lunch which costs $7.38. He has one $10 bill, 3 quarters, and 4 pennies. What should he offer to get back the fewest coins?

### Solution

If Rajat offers just the $10 bill, his change would be $10 − $7.38, or $2.62. He would likely receive one $2 coin, 2 quarters, 1 dime, and 2 pennies, that is, 6 coins.

If Rajat offers the $10 bill and 2 quarters, his change would be $10.50 − $7.38, or $3.12. He would likely receive one $2 coin, one $1 coin, 1 dime, and 2 pennies, that is, 5 coins.

If Rajat offers the $10 bill, 2 quarters, and 3 pennies, his change would be $10.53 − $7.38, or $3.15. He would likely receive one $2 coin, one $1 coin, 1 dime, and 1 nickel, that is, 4 coins.

Therefore, offering the $10 bill, 2 quarters, and 3 pennies results in the change with fewest coins.

## Practise

**1.** You have no small coin change. What coins would you get back if you offered each amount for a bill of $5.42?

**a)** $6 **b)** $7 **c)** $10 **d)** $20

**2.** What coins would you get back if you offered each amount for a bill of 28¢?

**a)** 30¢ **b)** 35¢ **c)** 50¢ **d)** $1

**3.** Select the two best amounts to offer for each bill amount to get back the fewest coins.

| | Bill Amount | Possible Amounts to Offer | | | |
|---|---|---|---|---|---|
| **a)** | $4.12 | $5.00 | $4.92 | $5.12 | $4.15 |
| **b)** | $4.51 | $4.75 | $5.00 | $5.51 | $5.01 |
| **c)** | $7.23 | $8.00 | $7.75 | $7.25 | $7.28 |

**4.** You have the following money available for each purchase below.

What would you offer to get back the fewest coins from each purchase?

**a)** $6.38 **b)** $18.16 **c)** $24.29 **d)** $22.58

**5.** Consider Rajat's situation in the Example. Why might Rajat choose to pay with just the $10 bill? Why might he choose to pay with the $10 bill, 2 quarters, and 3 pennies?

### Skills Check — Mental Math

Determine without the use of a calculator.

**a)** 10% × $12 **b)** 10% × $40 **c)** 10% × $7 **d)** 10% of $8.50

**e)** 5% × $12 **f)** 5% × $40 **g)** 5% × $7 **h)** 5% of $8.50

## 4.3 Taxes and Total Cost

### Explore

What is the price of the most expensive item subject to PST and GST that you can buy with $10?

### Develop

#### Example

Howard is buying a package of VCR tapes for $19.95. Both 7% GST and 8% PST will be added to the price.

**a)** Estimate the total cost to determine how much money Howard should get out to pay for his purchase.

**b)** Calculate the change Howard will receive.

#### Solution

**a)** $(7\% + 8\%)$ of $\$19.95 \doteq 15\% \times 20$ — 15% of the selling price

$= 0.15 \times 20$ — $0.10 \times 20 = 2$, and 0.05 is half as much, so $2 + 1 = 3$

$= 3$

$20 + 3 = 23$ — selling price + taxes

Howard should get out $23 to pay for his purchase.

**b)** *Step 1* Calculate the GST. 

$7\%$ of $\$19.95 = 0.07 \times 19.95$ — The cash register calculates the taxes separately.

$\doteq 1.40$

*Step 2* Calculate the PST.

8% of \$19.95 $= 0.08 \times 19.95$

$\doteq 1.60$

*Step 3* Calculate the total cost.

$19.95 + 1.40 + 1.60 = 22.95$ — selling price + GST + PST

*Step 4* Calculate the change.

$23 - 22.95 = 0.05$ — amount offered − total cost

The cashier gives Howard 5¢ change.

## Practise

**1.** Calculate the total taxes (GST and PST) on each purchase.

**a)** \$500 **b)** \$15 **c)** \$40 **d)** \$70

WS **2.** Each item purchased is subject to both taxes. Complete the table. When estimating, round to numbers that are easy to multiply.

| | Item | Cost | Estimate of Total Taxes | Estimate of Total Cost | Actual Total Taxes | Actual Total Cost | Cash Tendered | Change Received |
|---|---|---|---|---|---|---|---|---|
| **a)** | inline skates | \$99.50 | | | | | \$120.00 | |
| **b)** | helmet | \$39.95 | | | | | \$50.05 | |
| **c)** | clock radio | \$19.98 | | | | | \$25.00 | |
| **d)** | desk lamp | \$16.29 | | | | | \$20.03 | |

**3.** Steve is starting his own business. His children saved \$150 to buy him an answering machine priced at \$119.37.

**a)** Determine the total cost of the answering machine, including both taxes.

**b)** How much money do Steve's children have left to buy gift wrapping and a card?

WS **4.** You need to purchase the following school supplies. Each item is subject to both taxes.

a rollerball pen at \$2.08
an eraser at \$1.96
a package of highlighters at \$3.22
a 10-pack of pencils at \$1.43
a pad of paper at \$2.89

**a)** Estimate the total cost before taxes, the total taxes, and the total cost after taxes.

**b)** Calculate the total cost before taxes, the total taxes, and the total cost after taxes.

**c)** You have \$10 to spend on school supplies. How much more do you need?

5. You have seen that unless you need to know the taxes separately, you do not need to find 7% and 8%—just 15%. Similarly, if you do not need to know the amount of the taxes, you do not need to find 15% and add it to the price. You can just find 115% of the price.
   **a)** Find 115% of the price to determine the total cost of a $12.79 purchase that is subject to both taxes. Then find the change from $20.
   **b)** Explain why finding 115% of the price gives the total cost, including taxes.
   **c)** What percent would you use to determine the total cost in one step if only PST applied? Explain.
   **d)** What percent would you use to determine the total cost in one step if only GST applied? Explain.

6. **a)** Find the change from $30.36 for a purchase of $23.79 plus both taxes.
   **b)** What method did you use? Why?

7. You have $10 and you want to buy a frame for a photograph. The frame sells for $8.49 and is subject to PST and GST. Do you have enough money? Explain.

8. Restaurant bills of $4 or less are subject to GST only. Those more than $4 are subject to both taxes.
   **a)** What is the total cost of a slice of pizza and a drink for $3.99?
   **b)** What is the total cost of two hamburgers for $3.89 each, two orders of fries for $1.49 each, and two soft drinks for $1.29 each?
   **c)** Two friends each order a chicken dinner with a drink for $3.89. What would the total savings be if they had two bills rather than one?

**Skills Check** **Estimation**

Estimate.

**a)** $40 – $36.22 **b)** $59 + $63 + $138 **c)** 8 pens at $1.19 each
**d)** $25 – $19.17 **e)** $33 + $88 + $92 **f)** 7 apples at 27¢ each

# 4.4 Discounts and Sale Prices

A **discount** is a reduction in price. When discounts for goods and services are offered, the rate of the discount is usually advertised as a percent or a fraction of the regular, or list, price.

## Explore

A designer sweatshirt with a regular price of \$99 is on sale for 25% off. For a one-day sale, it is reduced by an additional 15% of the first sale price. Is the total discount 40%? Explain.

## Develop

### Example 1

Gwen is purchasing a tennis racket during a summer sale. It is advertised at 40% off the regular price of \$165.99.

**a)** Estimate the sale price.

**b)** Calculate the sale price.

### Solution

**a)** *Step 1* Estimate the discount.

$$\begin{aligned} 40\% \text{ of } \$165.99 &\doteq 0.40 \times 170 \\ &= 68 \end{aligned}$$

$4 = 2 + 2$, so $2 \times 170 = 340$, and $2 \times 340 = 680$
since $4 \times 170 = 680$, $0.40 \times 170 = 68$

*Step 2* Estimates the sale price.

$$\begin{aligned} \text{regular price} - \text{discount} &\doteq 170 - 68 \\ &= 102 \end{aligned}$$

The sale price is about \$102.

**b) Method 1** 

*Step 1* Calculate the discount.

$$\begin{aligned} \text{discount} &= 40\% \text{ of } \$165.99 \\ &= 0.4 \times 165.99 \\ &\doteq 66.40 \end{aligned}$$

*Step 2* Calculate the sale price.

$$\begin{aligned}\text{sale price} &= \text{regular price} - \text{discount}\\ &= 165.99 - 66.40\\ &= 99.60\end{aligned}$$

The sale price of the tennis racket is exactly \$99.60.

**Method 2**

*Step 1* Calculate the sale price percent.

$$\begin{aligned}\text{sale price percent} &= 100\% - 40\%\\ &= 60\%\end{aligned}$$

*Step 2* Calculate the sale price.

$$\begin{aligned}\text{sale price} &= 60\% \text{ of } \$165.99\\ &= 0.60 \times 165.99\\ &\doteq 99.60\end{aligned}$$

The sale price of the tennis racket is exactly \$99.60.

**1.** Which method in Example 1 do you prefer? Why?

**2.** Complete each statement.

**a)** If the discount is 20% of the regular price, then the sale price is ___% of the regular price.

**b)** If the discount is 30% of the regular price, then the sale price is ___% of the regular price.

**c)** If the discount is 15% of the regular price, then the sale price is ___% of the regular price.

**3.** Without first determining the discount, estimate the sale price of each item on sale at 30% off. When estimating, round to numbers that are easy to multiply.

**a)** regular price \$95.98 **b)** regular price \$28.75

**4.** Without first determining the discount, calculate the sale price of each item on sale at 15% off.

**a)** regular price \$79.98 **b)** regular price \$34.75

## Example 2

A carpet cleaning service advertises a January sale. It offers $\frac{1}{3}$ off its regular prices. The minimum charge is regularly \$39.99.

Calculate the minimum charge during the $\frac{1}{3}$ off sale.

## Solution 

**Method 1**

*Step 1* Calculate the discount.

$\frac{1}{3} \times 39.99 = 13.33$

*Step 2* Calculate the sale price.

$$\begin{aligned}\text{regular price} - \text{discount} &= 39.99 - 13.33\\ &= 26.66\end{aligned}$$

The minimum price during the $\frac{1}{3}$ off sale is \$26.66.

**Method 2**

*Step 1* Calculate the discount as a fraction.

$1 - \frac{1}{3} = \frac{2}{3}$

*Step 2* Calculate the sale price.

$\frac{2}{3} \times 39.99 = 26.66$

The minimum charge during the $\frac{1}{3}$ off sale is \$26.66.

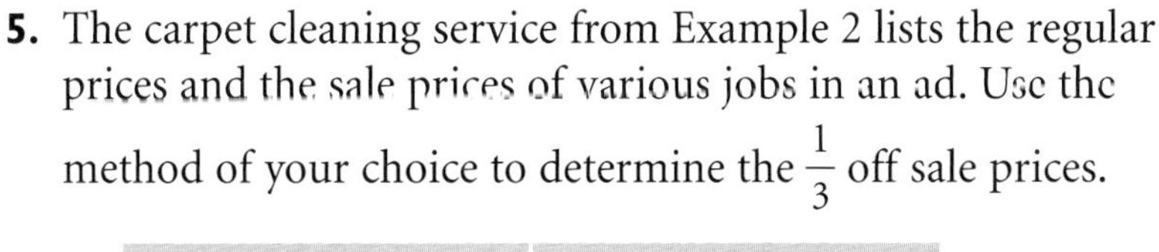

**5.** The carpet cleaning service from Example 2 lists the regular prices and the sale prices of various jobs in an ad. Use the method of your choice to determine the $\frac{1}{3}$ off sale prices.

| | Regular Price | Sale Price |
|---|---|---|
| **a)** | \$49.99 | |
| **b)** | \$59.99 | |
| **c)** | \$69.99 | |
| **d)** | \$79.99 | |
| **e)** | \$89.99 | |

## Practise

**6.** A business supply store offers students with a valid student card 30% off all stationery purchases. How much will Ay-Ling pay before taxes for \$14.63 worth of stationery products by showing her student card?

**7.** One store has shoes with a regular price of \$69.50 on sale at 10% off. Another store has the same shoes with a regular price of \$85 on sale at 25% off. Predict and then calculate which store has the better sale price.

WS **8.** Complete the table. When estimating, round to numbers that are easy to multiply.

| | Item | Regular Price | Discount Rate | Estimate of Sale Price | Actual Sale Price |
|---|---|---|---|---|---|
| **a)** | inkjet printer | \$234.99 | 10% | | |
| **b)** | mattress set | \$649.00 | 15% | | |
| **c)** | DVD/CD player | \$359.98 | 20% | | |
| **d)** | digital camera | \$499.00 | 20% | | |
| **e)** | watch | \$55.00 | 15% | | |
| **f)** | book | \$26.50 | 20% | | |

**9.** Calculate the sale price of each item.

**a)** a \$79.99 jacket on sale at 40% off

**b)** a \$185.49 maple tree on sale for $\frac{1}{3}$ off

**c)** a \$3999 notebook computer on sale at 10% off

**10.** During a "Scratch-and-Save" sale, Ron saved 10% on a \$34.99 coffee maker for his mother for Mother's Day. His parents both enjoyed the coffee maker. Another "Scratch and Save" sale was held before Father's Day. Ron saved 25% on a \$33.49 coffee grinder for his dad. What was the sale price of each item?

**11.** A clothing store had a "Buy Two Items and Get the Lower-Priced Item at Half-Price" sale. Janice purchased a top with a regular price of \$39.99 and pants with a regular price of \$44.99. What was her total bill before taxes?

**12.** Charlotta is planning to purchase a scanner. She learns that a store is having a "Going Out of Business" sale. It is offering $\frac{2}{3}$ off all merchandise, including a scanner that interests her.

**a)** What is the sale price of the scanner if its regular price is \$549.99?

**b)** What concerns might Charlotta have about purchasing a computer item from this store?

**13.** A discount store offers all its merchandise at 25% off the manufacturers' list prices. When merchandise has been in the store one month, it is reduced by 15% of the first discounted price. When it has been in the store two months, it is reduced by 20% of the second discounted price. Use an item with a list price of \$100 to show that the discount after two months in the store is not equal to a 60% (the sum of 25%, 15%, and 20%) discount. Explain your results.

# 4.5 Career Focus: Sales and Merchandising Clerk

Sandy is a clerk in a hardware store. One task that she enjoys is creating signs for end-of-aisle displays. The policy of the store is to have a sign for sale items on end-of-aisle displays featuring this information:

- the name of the item
- the sale price
- the regular price
- the discount rate
- the stock item number

The item name and the sale price always appear in the largest print.

Sandy created this sign using computer software.

**1.** Create a sign that meets the specifications given above for each sale. Use a word processor or other computer software, or draw by hand.

**a)** bicycle helmets
regular price $24.99
discount rate 15% off
small #10-365 medium #10-366 large #10-367

**b)** watering cans
regular price $2.49
discount rate half price
#33-3487

**c)** garden shovels
regular price $9.79
discount rate 10% off
#33-4987

**2.** On your sign, what is the ratio of the height of the largest letters to the height of the smallest letters?

**3.** Another job that Sandy enjoys is pricing new merchandise. The selling price is the cost to the store plus a markup. The **markup** rate is a percent of the cost to the store. Stores mark up their costs to pay their overhead, such as building maintenance, rent or taxes, utilities, salaries and benefits for employees and, of course, to make a profit.

The following is a list of new merchandise at the hardware store. The markup code is explained below the list.

| Item | Stock Item # | Cost per Item | Markup Code |
|---|---|---|---|
| hammer | 73-2318 | $9.24 | R |
| sandpaper | 73-8719 | $0.19 | R |
| measuring tape | 73-3240 | $2.12 | S |
| door stop | 20-7493 | $0.78 | R |
| window cleaner | 42-3938 | $1.23 | T |
| picture wire | 22-4829 | $0.84 | R |
| pack of sponges | 42-0820 | $0.47 | T |
| air freshener | 42-2767 | $1.02 | T |
| plant food | 33-5938 | $2.24 | T |
| broom | 42-5720 | $1.35 | S |
| package of tacks | 22-5934 | $0.28 | S |
| Markup Code (% of cost): R 35% S 50% T 65% | | | |

Sandy determines the selling price of the new model of hammers by one of two methods.

**Method 1**
Find 35% of $9.24, add the amount to $9.24, and round up to the next 9¢.

**Method 2**
Find 135% of $9.24 and round up to the next 9¢.

**a)** Find the selling price of the new hammer using both methods. Which do you prefer? Why?
**b)** Why do you think the selling price is rounded up to the next 9¢?

**c)** Use the method of your choice. Find the selling prices for the other items of new merchandise listed above.

**4.** Sandy also works on the cash register. She makes change for customers' purchases. Determine the bills and coins she should give to the customer of each purchase for the amounts tendered.

**a)** total price of $13.78, when $15 tendered
**b)** total price of $20.67, when $22.17 tendered
**c)** total price of $26.28, when $31.30 tendered
**d)** total price of $32.52, when $40 tendered
**e)** total price of $8.93, when $10.03 tendered

# 4.6 Sale Prices, Taxes, and Total Cost

## Explore

You are purchasing two items on sale at 20% off their regular prices of \$11.95 and \$7.89. Both items are subject to PST and GST. Will \$20 be enough to pay for them? Explain.

## Develop

### Example

Emily is buying an item on sale at 30% off its regular price of \$58.89. It is subject to both PST and GST. Estimate and calculate the total cost.

### Solution

**Method 1**

**Estimate.**

Discount

$30\%$ of $\$58.89 \doteq 0.30 \times 60$
$= 18$

Use numbers that are easy to multiply.

30% of regular price

Sale price

$60 - 18 = 42$

regular price − discount

Taxes

$15\%$ of $\$42 = 0.15 \times 42$
$= 6.30$

15% of sale price

$10\% \rightarrow 4.2 \quad 5\% \rightarrow 2.1 \quad 4.2 + 2.1 = 6.3$

Total cost

$42 + 6.3 = 48.30$

sale price + taxes

The total cost is approximately \$48.30.

**Calculate.** 

Discount

$30\%$ of $\$58.89 = 0.30 \times 58.89$
$\doteq 17.67$

30% of regular price

Sale price

$58.89 - 17.67 = 41.22$

regular price − discount

Taxes

$15\%$ of $\$41.22 = 0.15 \times 41.22$
$\doteq 6.18$

15% of sale price

Total cost

$41.22 + 6.18 = 47.40$

sale price + taxes

The total cost is \$47.40.

**Method 2**

**Estimate.**

Sale price

70% of \$58.89 $\doteq 0.70 \times 60$ — Discount is 30%, so use 70% for sale price.

$= 42$ — Use numbers that are easy to multiply.

Total cost including taxes

115% of \$42 $= 1.15 \times 42$ — 100% + 8% PST + 7% GST is 115%.

$= 48.3$ — 100% → 42 10% → 4.2 5% → 2.1 42 + 4.2 + 2.1 = 48.3

The total cost is approximately \$48.30.

**Calculate.** 

Sale price

70% of \$58.89 $= 0.70 \times 58.89$

$\doteq 41.22$

Total cost including taxes

115% of \$41.22 $= 1.15 \times 41.22$

$\doteq 47.40$

The total cost is \$47.40.

Which method do you prefer? Would you use a different method?

## Practise

**1.** Complete the table below. All items are subject to both taxes. When estimating, round to numbers that are easy to multiply.

| | Item | Regular Price | Discount Rate | Estimate of Sale Price | Estimate of Total Cost | Actual Sale Price | Actual Total Cost |
|---|---|---|---|---|---|---|---|
| **a)** | binder | \$5.89 | 10% | | | | |
| **b)** | calculator | \$15.79 | 15% | | | | |
| **c)** | luggage | \$42.95 | 20% | | | | |
| **d)** | shirt | \$49.99 | $\frac{1}{3}$ | | | | |
| **e)** | bookcase | \$59.98 | 15% | | | | |

**2.** Calculate the cost of dry cleaning a jacket. The regular price is \$18.95, but there is now 15% off the service. Dry cleaning is subject to GST, but not PST.

**3.** Enid buys a cordless mouse on sale at 25% off the regular price of \$79.95 and a pack of formatted disks on sale at 25% off the regular price of \$8.50. Both items are subject to both taxes. Will \$75 pay for the two purchases? Explain your answer by showing your work.

**4.** Mia purchases a dress with a regular price of \$100 on sale at 15% off. The purchase is subject to GST and PST. Why is her total bill less than \$100?

# 4.7 Putting It All Together: Discounts, Taxes, and Total Cost

**A** You and a friend are going camping at the end of August. You want to spend five days in a provincial park. You need to select
- a park to go to
- camping gear to buy, at least some of which should be on sale as you are on a budget
- food to buy for the trip

Apply what you have learned in this chapter to find a reasonable estimate of the cost of the trip. Show all of your work including discounts received and taxes paid. You do not need to consider transportation costs at this stage of planning.

## 4.8 Chapter Review

**For the situations in questions 1 to 4, answer the following.**
**a) What amount offered should be entered in the cash register?**
**b) What change does the cash register calculate?**
**c) What bills and coins should the customer receive based on using the fewest coins possible?**
**d) Check your answer to part c) by counting out the change.**

**1.** Punam gives the cashier a $10 bill to pay for a $6.81 purchase.

**2.** Frederick gives the cashier a $20 bill, a nickel, and 2 pennies to pay for items totalling $17.32.

**3.** Aaron offers a $10 bill and a $5 bill for his purchase of $13.79 worth of groceries.

**4.** Tina gives the cashier a $5 bill, 2 dimes, and a penny to pay for a $3.71 purchase.

**5.** Angelica purchases a few items from the drugstore. The total bill is $4.27. She has one $5 bill, 2 quarters, 1 nickel, and 4 pennies. What should she give the cashier to get back the fewest coins in the change?

**6.** Tony purchases a loaf of bread for $1.79 and a jar of jam for $3.29. He has one $10 bill, 2 quarters, 1 dime, 3 nickels, and 2 pennies. What should he give the cashier to get back the fewest coins in the change?

**7.** Basmattie is purchasing a new frying pan for $47.99 plus GST and PST.

**a)** Estimate and then calculate the total cost.
**b)** What percent(s) did you calculate and why?

WS **8.** Each purchase listed below is subject to both taxes. Complete the table. When estimating, round to numbers that are easy to multiply.

| | Item | Cost | Estimate of Total Taxes | Estimate of Total Cost | Actual Total Taxes | Actual Total Cost | Cash Tendered | Change Received |
|---|---|---|---|---|---|---|---|---|
| **a)** | pants | $47.95 | | | | | $60.15 | |
| **b)** | shirt | $29.95 | | | | | $40.00 | |
| **c)** | socks | $3.29 | | | | | $4.03 | |
| **d)** | shoes | $58.89 | | | | | $70.00 | |

**9.** What is the price of the most expensive item subject to PST and GST that you can buy with $25?

WS **10.** Complete the table below. When estimating, round to numbers that are easy to multiply.

| | Item | Regular Price | Discount Rate | Estimate of Sale Price | Actual Sale Price |
|---|---|---|---|---|---|
| **a)** | hockey stick | $34.99 | 10% | | |
| **b)** | putter | $44.99 | 15% | | |
| **c)** | baseball glove | $54.96 | 20% | | |

**11.** One store has put a CD with a regular price of $22.95 on sale at 10% off. A discount store has the same CD for $20.50. Predict and then calculate which store has the better buy.

**12.** A store pays $26.25 each for a line of dresses. It marks them up 100%, and then immediately puts them on "sale" at 25% off.

**a)** What is the "sale" price of a dress?

**b)** Since the store sells the dresses at the "sale" price only, what is the actual markup?

WS **13.** Complete the table below. All items are subject to both taxes. When estimating, round to numbers that are easy to multiply.

| | Item | Regular Price | Discount Rate | Estimate of Sale Price | Estimate of Total Cost | Actual Sale Price | Actual Total Cost |
|---|---|---|---|---|---|---|---|
| **a)** | dartboard | $25.99 | 10% | | | | |
| **b)** | table tennis table | $269 | $\frac{1}{3}$ | | | | |
| **c)** | litre of paint | $38.95 | 20% | | | | |

**14.** Without performing any calculations, answer the following.

A purchase subject to GST and PST with a regular price of $75 is on sale at 15% off. Is the total cost more than, less than, or equal to $75? Explain.

**15.** Andrew is buying a friend a birthday gift. The regular price is $39.95, but the item is on sale at 10% off. It is subject to both taxes. How much change does he get back if he offers two $20 bills and a $5 bill?

CHAPTER

# 5 Buying Decisions

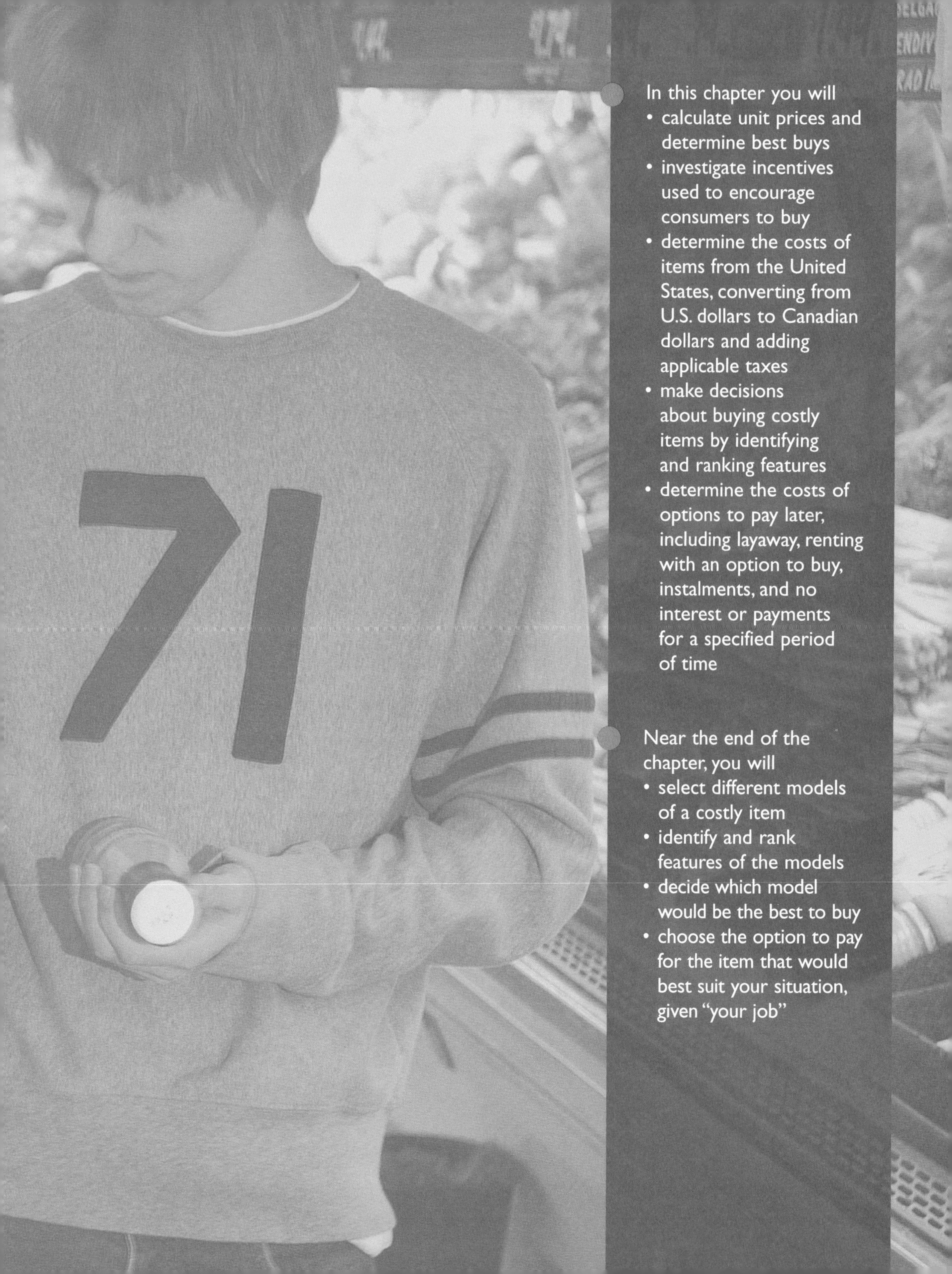

In this chapter you will

- calculate unit prices and determine best buys
- investigate incentives used to encourage consumers to buy
- determine the costs of items from the United States, converting from U.S. dollars to Canadian dollars and adding applicable taxes
- make decisions about buying costly items by identifying and ranking features
- determine the costs of options to pay later, including layaway, renting with an option to buy, instalments, and no interest or payments for a specified period of time

Near the end of the chapter, you will

- select different models of a costly item
- identify and rank features of the models
- decide which model would be the best to buy
- choose the option to pay for the item that would best suit your situation, given "your job"

# 5.1 The Best Buy

Most grocery store shelves have stickers that state the **unit prices** of the items on the shelves. Here's an example: \$0.445/100 g, which means 44.5¢ per 100 g.

The price is in dollars or cents. The unit varies. Here are some examples:

- kilogram
- gram
- 100 g
- litre
- millilitre
- 100 mL
- package

For each unit listed above, name some typical products that are sold that way.

## Explore

A package of 2 AA batteries sells for \$3.99. A package of 8 AA batteries sells for \$9.99. Under what circumstances might you buy the 2-pack? the 8-pack?

## Develop

### Example 1

Paper towels are sold in a 2-roll package for \$2.49 and a 12-roll package for \$12.99.

**a)** Which package has the lower unit price?

**b)** How much would you save by buying a 12-roll package rather than six 2-roll packages?

**c)** When deciding which package size is the better buy for you, what should you consider in addition to unit price?

### Solution 

**a)** *Step 1* Calculate the unit price per roll for each package size by dividing the cost by the number of rolls.

unit price per roll for a 2-roll package
$2.49 \div 2 = 1.245$ Do not round.

unit price per roll for a 12-roll package
$12.99 \div 12 = 1.0825$

The unit price per roll of a 2-roll package is \$1.245 and of a 12-roll package, \$1.0825.

*Step 2* Compare the unit prices.

\$1.0825 is less than \$1.245.
The 12-roll package has the lower unit price.

**b)** *Step 1* Calculate the cost of six 2-roll packages.
$6 \times 2.49 = 14.94$

*Step 2* Calculate how much more that is than the cost of a 12-roll package, $12.99.
$14.94 - 12.99 = 1.95$
The savings is $1.95 for a 12-roll package.

**c)** Do I need 12 rolls of paper towels?
Do I have $12.99 to spend on one item?
Do I have room to store 12 rolls of paper towels at a time?

## Example 2

"Mr. Potato" potato chips are sold in three sizes:
515 g for $3.99     270 g for $2.49     410 g for $3.49

**a)** Calculate the unit price per gram for each bag size.

**b)** Calculate the unit price per 100 g for each bag size.

**c)** Why might the unit price be given per 100 g rather than per gram?

**d)** Which size of bag has the lowest unit price?

**e)** When deciding which of the three bag sizes is the best buy, what are some considerations other than unit price?

### Solution

**a)** 515 g bag/gram: $3.99 \div 515 = 0.0077$    Do not round.
270 g bag/gram: $2.49 \div 270 = 0.009\ 22$
410 g bag/gram: $3.49 \div 410 = 0.0085$

The unit prices are $0.0077/g for a 515 g bag, $0.009 22/g for a 270 g bag, and 0.0085/g for a 410 g bag.

**b)** Multiply each answer in part a) by 100.
515 g bag/100 g: $0.0077 \times 100 = 0.77$
270 g bag/100 g: $0.009\ 22 \times 100 = 0.922$
410 g bag/100 g: $0.0085 \times 100 = 0.85$

The unit prices are $0.77/100 g for a 515 g bag, $0.922/100 g for a 270 g bag, and $0.85/100 g for a 410 g bag.

**c)** The numbers are not as small, and with fewer decimal places they are easier to read.

**d)** Since $0.77/100 g is the least amount, the 515 g bag has the lowest unit price.

**e)** Since potato chips are perishable, you must consider whether all of the 515 g bag will be consumed before the contents go stale.

## Practise

1. Calculate the unit price for the following in
   a) dollars per kilogram: 2 kg of carrots for $2.59
   b) dollars per gram: 425 g box of cereal for $3.49
   c) dollars per 100 g: 425 g box of cereal for $3.49
   d) dollars per bag: 12 bags of popcorn for $5.99
   e) dollars per can: 24 cans of pop for $6.49

2. Consider these three ways of selling milk:
   4 L for $3.49    2 L for $2.89    1 L for $1.79
   a) Calculate the unit price per litre for each size.
   b) Which has the lowest unit price?
   c) What else should you consider before selecting the best size to buy?

3. Lalena examined the prices of two packages of lean ground beef:
   $10.22 for 1.950 kg    $3.40 for 0.620 kg
   a) Calculate the unit price per kilogram for each package of lean ground beef.
   b) Why are the unit prices for the same cut of beef not the same?
   c) How might the number of people you plan to feed affect your purchase decision?

4. Consider these two ways of selling apple juice:
   package of eight 200 mL boxes for $2.59    1.36 L can for $1.29
   a) How many litres are in a package of eight 200 mL boxes?
   b) What is the unit price per litre for each way?
   c) Which is the lower unit price?
   d) What other than unit price would affect your purchase decision?

5. A large amusement park has adult season passes for $92.95. An adult day pass is $44.99. How many times would you have to visit the amusement park to benefit from the savings offered by a season pass? Explain your answer by showing your work.

6. A zoo has annual memberships for $70. Of that, $45 is for admission and $25 is for parking. A single admission price is $15 and parking is $6. How many times would you have to go to the zoo to realize the savings offered by a membership?

### Skills Check — Percent of a Number

Determine the total cost, including 8% PST and 7% GST, of each item.

a) a shirt for $25.75
b) a pair of shoes for $69.99
c) a CD for $20.95
d) a jacket for $87.49
e) a car for $27 500
f) a radio for $39.97

# 5.2 Incentives to Buy

## Explore

What are some practices that stores and manufacturers follow to encourage people to buy from them or buy their products?

## Develop

1. Some businesses distribute **discount coupons**, or ads of sale prices where the ads must be handed in to get the sale prices. Manufacturers also distribute discount coupons for particular items that can be redeemed at a variety of stores. List three ways that coupons are distributed.

2. A large hardware chain offers **store money**, or money that can be used towards future purchases at its stores. For a minimum purchase of $2.50 and payment in cash or by debit card, you get store money. The amount is 2.5% of the total before taxes. The store money is rounded to the nearest 5¢. For example, 2.5% of $35 is $0.875, which rounds up to $0.90, or 90¢.

   Calculate the amount of store money you would receive if you paid cash for each of these items.

   **a)** a halogen headlight at $15.97
   **b)** a slide compound mitre saw at $749
   **c)** a 12 V cordless drill at $199
   **d)** an air hockey game at $399
   **e)** a lacrosse stick at $59

3. Some department stores issue **points cards**. If you present your points card when paying for items, you earn 50 base points for every before-taxes dollar spent.

   How many points would you earn on each item listed here?

   **a)** a $20 shirt
   **b)** a $120 bicycle
   **c)** a $15 video
   **d)** a $200 bookcase

4. At 50 base points for every before-taxes dollar spent, what value of merchandise would you need to purchase to earn 10 000 points?

5. Look through a catalogue that allows you to use points when you make purchases. Select an item to buy using between 100 000 and 500 000 points. If your point total is insufficient to buy the item, can you make a money payment in combination with the points you have? If so, how many points and how much money?

6. A large drugstore chain has a program that uses points cards. Use brochures or the Internet to investigate a program like this and answer the following questions.

Access to the Web site for this points program can be gained through the *Mathematics for Everyday Life 11* page of irwinpublishing.com/students.

**a)** What is the cost of membership in the program?
**b)** How many points are earned for every before-taxes dollar spent?
**c)** List five items on which points are not awarded and may not be redeemed.
**d)** Some items offer Bonus Points. How does a customer know whether bonus points apply to an item?
**e)** Complete the following table that explains how the program works.

| Points Required | Discount Reward | Reward Value Up To |
|---|---|---|
| 3 000 | 20% off | |
| 6 000 | | $10 |
| | 55% off | $25 |
| 20 000 | | $40 |
| 26 000 | 85% off | |
| | 100% off | |

**f)** How are points redeemed?

## Practise

7. Some points programs have many sponsors. Use brochures or the Internet to investigate such a program and answer the following questions.

Access to the Web site for this points program can be gained through the *Mathematics for Everyday Life 11* page of irwinpublishing.com/students.

**a)** List five sponsors of the program.
**b)** Does the list of sponsors change over time? If so, name a new sponsor.
**c)** List three ways in which points can be redeemed.

8. Credit cards allow you to make purchases and pay for them later. You will investigate making purchases using credit cards in Chapter 9. Many large retailers offer their own credit cards with incentives to encourage you to use their cards.

   The hardware chain with store money for cash payments offers a credit card with incentives. With the card you receive 20% more store money than you do when paying cash.

   **a)** If you get 2.5% of the total in store money for paying cash, what percent of the total is given in store money when you use the card?

   **b)** You are not handed the store money when you use the credit card. How would you accumulate the store money?

9. Use brochures or the Internet to investigate the credit card incentive described in question 8 and answer the following questions.

   **a)** How do you redeem the store money when you don't have it in hand?

   **b)** Does an annual fee or transaction fee apply to this card? If so, what is it?

Access to the Web site for this incentive program can be gained through the *Mathematics for Everyday Life 11* page of irwinpublishing.com/students.

10. Another credit card offered by a store with a points card has incentives that entitle purchasers to 25 bonus points for every before-taxes dollar spent using the credit card.

    **a)** If you have a points card, you earn 50 base points for every before-taxes dollar spent no matter how you pay. How many points would you earn if you bought a $70 coat and didn't use the credit card?

    **b)** How many points would you earn if you used the credit card to buy the $70 coat?

11. A large grocery chain offers a credit card with incentives. Use brochures or the Internet to investigate this credit card incentive and answer the following questions.

    **a)** Is there an annual fee or transaction fee for using this card?

    **b)** How do you collect points with this credit card?

    **c)** How do you redeem the points?

Access to the Web site for this incentive program can be gained through the *Mathematics for Everyday Life 11* page of irwinpublishing.com/students.

## Skills Check **Mental Math**

Use mental math to multiply each price by 5.

| | | | |
|---|---|---|---|
| **a)** $28 | **b)** $42 | **c)** $66 | **d)** $89 |
| **e)** $103 | **f)** $124 | **g)** $218 | **h)** $507 |

# 5.3 Cross-Border Shopping

With increased access to goods from the United States through travel and the Internet, a Canadian consumer would find it helpful to convert from U.S. dollars to Canadian dollars.

The conversion factors change daily because the exchange rate varies daily. For example, on one day the U.S. dollar to Cdn dollar conversion factor was $1.00 U.S. = $1.545 00 Cdn.

Access to a Web site for current conversion factors can be gained through the *Mathematics for Everyday Life 11* page of irwinpublishing.com/students.

## Explore

**Estimate the price in Canadian dollars of a skateboard advertised on the Internet for $175 U.S. What other costs might be involved?**

## Develop

### Example

While travelling in the United States, Jocelyn sees a watch for $44.95. Estimate and calculate the price in Canadian dollars using the conversion factor given above.

### Solution

**Estimate.**

price in U.S. dollars times conversion factor $= 44.95 \times 1.545$

$\doteq 46 \times 1.5$ (Half of 46 is easier than half of 45.)

$= 46 + 23$ (46 + half of 46)

$= 69$

The price of the watch in Canadian dollars is about $69.

**Calculate.** 

price in U.S. dollars times conversion factor $= 44.95 \times 1.545$

$\doteq 69.45$

The price of the watch in Canadian dollars is exactly $69.45.

## Practise

**Use the conversion factor $1.00 U.S. = $1.545 00 Cdn.**

**1.** Estimate and calculate each price in Canadian dollars.

**a)** $15.95 U.S. **b)** $4.88 U.S. **c)** $29.88 U.S. **d)** $279 U.S.

**2.** The prices shown were taken from U.S. and Canadian flyers for identical items. In which country is each item cheaper?

| | Item | U.S. Price in U.S $ | Canadian Price in Cdn $ |
|---|---|---|---|
| **a)** | 12-pack pop | $1.99 | $2.99 |
| **b)** | phone | $19.99 | $25.99 |
| **c)** | 4-pack AA batteries | $2.50 | $3.95 |
| **d)** | toothpaste | $1.49 | $1.99 |

**For the scenarios in questions 3 to 6, use the following information and answer a) to d).**

If you purchase items in the United States and bring them into Canada, the items you purchase may be subject to taxes. You pay the taxes at Canada Customs as you cross the border into Canada.

- The price of the item is converted to Canadian dollars.
- You pay no taxes if the price in Canadian dollars is less than
  - $50 and you have been in the United States for 24 hours
  - $200 and you have been in the United States for 48 hours
  - $750 and you have been in the United States for 7 days

  These are called **personal exemptions**.
- If you have been in the United States for 24 hours, and you make purchases of more than $50 Cdn, you pay PST and GST on the total amount.
- If you have been in the United States for 48 hours, and you have made purchases of more than $200 Cdn, you pay PST and GST on the amount exceeding $200.
- If you have been in the United States for 7 days, and you have made purchases of more than $750 Cdn, you pay PST and GST on the amount exceeding $750.

**a)** Estimate the cost in Canadian dollars.
**b)** Calculate the cost in Canadian dollars.
**c)** On what amount, if any, must PST and GST be paid?
**d)** What is the total PST and GST, if any, that must be paid?

**3.** Marita purchases a U.S.-made computer monitor for $125 U.S. during a three-day conference in Houston.

**4.** Paul buys a U.S.-made coat for $59 U.S. during an overnight visit to Buffalo.

**5.** Jill buys a U.S.-made cordless telephone for $65 U.S. while vacationing in Chicago over Saturday and Sunday.

**6.** Wilma buy a U.S.-made vacuum cleaner for $649 U.S. while staying with her cousin for a month in Pennsylvania.

# 5.4 Career Focus: Musician

Jenna loves music—she plays the piano and the guitar, and she sings. She works 20 hours a week at a music store where she is paid $8 per hour. Jenna also volunteers at a library and belongs to a band.

In the band that Jenna has with a group of friends, she sings lead vocals and plays the guitar. Ahmed plays percussion, Kevin plays keyboard, and Melanie plays bass. The band is called "Drive Shed" because it practises at Melanie's farm in the drive shed.

1. Jenna wants to purchase an acoustic guitar. She found three of interest to her in a U.S. catalogue.

   | | |
   |---|---|
   | guitar #1 | $1285 U.S. |
   | guitar #2 | $1069 U.S. |
   | guitar #3 | $649 U.S. |

   Determine the price of each guitar in Canadian dollars using the current conversion factor or $1.00 U.S. = $1.545 00 Cdn.

2. As a volunteer at a library, Jenna plays music and runs a singsong for preschoolers. She does this one morning a week because she wants to start a part-time business in which she performs at children's birthday parties. How would working with preschool children help her launch this business?

3. Jenna needs to develop a repertoire of songs for her volunteer work and her business. She decides to buy two children's music books. At the bookstore, she gets a 20% discount off their selling prices of $44.95 and $28.95. What is the total cost of the books, which are subject to GST, but not PST?

4. Jenna decides to make balloon figures as props for the parties. She plans to buy a kit with a How-to book, a pump, and 20 balloons for $9.99. She also plans to buy more balloons, either in packages of 100 for $10 or in packages of 144 for $12.99.
   **a)** What is the cost per balloon in the first package?
   **b)** What is the cost per balloon in the second package?
   **c)** Which package is the better buy?
   **d)** Jenna buys the kit and four packages of the "better buy" balloons at a store where she earns 50 points for every before-taxes dollar spent. How many points does she earn?

5. Publicity from her volunteer work at the library proves to be successful. Jenna is booked for two parties a week for the next eight weeks. She will earn $75 per party.

   **a)** How much will she earn each week for the parties?
   **b)** How much will she earn in the next eight weeks for the parties?
   **c)** How much will she earn in the next eight weeks for the parties and from her bookstore job?

6. Jenna's band, "Drive Shed," participated in a band competition.

   - Each band paid $25 to participate.
   - Each band performed two numbers.
   - All performances were recorded live to be released as a live concert CD selling for $15.
   - Tickets to attend the concert cost $5.
   - The competition prize was a free recording session worth $2000.
   - The recording studio charged the competition organizers only $1000 for the $2000 prize.

   **a)** Why would a recording studio charge the competition organizers only half the price?
   **b)** How many tickets would the competition organizers have to sell to pay for the prize?
   **c)** How much money would the competition organizers collect if they sold 850 CDs?

7. **a)** How many hours would Jenna need to work at the music store to pay for a $2000 recording session?
   **b)** How many 20-hour workweeks does this total represent?

8. "Drive Shed" won the competition and recorded a CD at the recording session. The band plans to perform throughout Ontario next summer. How might they use their CD to help fulfill their plans?

## 5.5 Deciding Which to Buy

### Explore

Assume you are shopping for a television. What features would you want to know about the models available? How would you use the information to help you decide which to buy?

### Develop

**1.** Here are some specifications of three personal watercraft models.

| Specification | Model A | Model B | Model C |
|---|---|---|---|
| displacement | 1176 cc | 1176 cc | 1131 cc |
| horsepower | 155 hp@7000 rpm | 155 hp@7000 rpm | 135 hp@6750 rpm |
| length | 3.16 m | 2.93 m | 3.85 m |
| width | 1.22 m | 1.15 m | 1.51 m |
| height | 1.13 m | 1.02 m | 1.14 m |
| dry weight | 354 kg | 306 kg | 370 kg |
| engine | 3 cylinder, 2-stroke | 3 cylinder, 2-stroke | 3 cylinder, 2-stroke |
| vehicle capacity | 240 kg | 160 kg | 300 kg |
| passengers | 1 to 3 | 1 or 2 | 1 to 4 |
| storage capacity | 90.5 L | 16 L | 441 L |
| fuel capacity | 70 L | 60 L | 70 L |
| price | $11 999 | $11 999 | $13 299 |

Identify the model(s) with

**a)** the least dry weight
**b)** the greatest fuel capacity
**c)** the greatest width
**d)** the least length

**2.** Which of the personal watercraft models listed in question 1 would you select in each of these cases?

**a)** There will never be more than two passengers and price is an issue.
**b)** Lots of storage capacity is the number-one requirement.
**c)** There will be three passengers and a strong horsepower is valued.
**d)** You are the buyer. Explain the reasons for your selection.

**3.** Where could you find information about other models of personal watercrafts?

4. When David decided to buy a car, he determined that these features were important to him:

| | | |
|---|---|---|
| • CD player | • fuel consumption | • one to three years old |
| • cost under $16 000 | • radial tires | • 6-cylinder engine |
| • colour | • four doors | • air conditioning |

Before he began shopping, he ranked the features in order of importance. The most important was cost under $16 000 and the least important was colour.

**a)** If you were buying a car, which of the features listed above would be your highest ranking? Why?

**b)** Would you have a higher ranking feature? If so, what is it and why would you rank it higher?

## Practise

5. Assume that you plan to purchase each of the following items. List at least four features of each and rank them in order of importance to you.

| | | |
|---|---|---|
| **a)** sofa | **b)** camera | **c)** refrigerator |
| **d)** winter coat | **e)** cruise vacation | **f)** Internet service provider |

6. Choose one of these items:

| | | |
|---|---|---|
| cell phone | DVD player | clothes washer |
| stereo system | guitar | item of your choice |

Using newspaper ads, flyers, catalogues, brochures, and/or the Internet, answer the questions that follow.

**a)** Identify at least four models of the item.

**b)** List at least six features for each model, including the price.

**c)** Rank the features in order of importance to you.

**d)** Identify which model you would buy and justify your decision.

Access to Web sites for the items in question 6 can be gained through the *Mathematics for Everyday Life 11* page of irwinpublishing.com/students.

### Skills Check — Mental Math

Use mental math to find the total cost of each of the following.

| | |
|---|---|
| **a)** 4 shirts at $18 each | **b)** 4 cell phones at $150 each |
| **c)** 4 airfares at $520 each | **d)** 4 tickets at $8.50 each |

Use mental math to find each individual amount.

| | |
|---|---|
| **e)** 4 people share $86 | **f)** 4 tires cost $296 |

# 5.6 Options to Pay—Layaway

In addition to paying the full amount at the time of purchase by cash, cheque, debit card, or credit card, some retailers offer other options to pay for goods. These include

- layaway
- renting with an option to buy
- instalments
- no interest or payments for a specified period of time

With a **layaway** purchase, a buyer puts down a **deposit**, often a percent of the price, including taxes. The remaining amount is paid when the item is picked up. No fees apply.

WS

## Develop

**1.** Joe sees the ideal necklace to give his friend for her birthday which is four months away. He puts 10% down and pays the remainder in four months. The necklace costs $159 plus PST and GST.

**a)** What is the total cost, including taxes?
**b)** How much of a deposit does he put down?
**c)** How much does he pay when he picks up the necklace?

**2.** Emily purchases a watch that costs $125 plus both taxes. She takes advantage of the layaway plan and puts 10% down.

**a)** What is the total cost of the watch, including taxes?
**b)** How much of a deposit does she put down?
**c)** How much does she pay when she picks up the watch?

**3.** Daniel purchases a bicycle that costs $280 plus both taxes. He puts 15% down and lays away the bike for 6 months.

**a)** What is the total cost of the bicycle, including taxes?
**b)** How much of a deposit does he put down?
**c)** How much does he pay when he picks up the bicycle?

## Practise

**4.** What do you consider to be the main disadvantages of layaway? Explain.

**5.** **Special orders**, where something is going to be brought in or made especially for you, work much like layaway. Carla ordered a roman window shade that cost $215 plus taxes. She paid 40% of the total cost when she placed the order and the remainder when she picked it up five weeks later. How much did she pay each time?

# 5.7 Renting with an Option to Buy

Rent-to-own stores offer another alternative. When **renting with an option to buy**, a consumer agrees to pay a monthly rental fee for a maximum rental period. If the item is rented for the maximum rental period, the consumer will own the item.

## Explore

Why might someone rent electronics, furniture, or appliances from a rent-to-own store rather than buy from a store? What other options might a person consider?

## Develop

**1.** A rent-to-own store has televisions with a cash price of $305. The monthly rental fee is $39. The maximum rental period is 18 months. The store offers free delivery, setup, and pickup, as well as parts and service for the rental period.

**a)** Determine the total cash price of a TV, including taxes.
**b)** Determine the total amount paid for a TV if rented for the entire 18 months.
**c)** How much more than the cash price is the rental price? Compare that to the cash price.

**2.** Jasmine rents the same television model as in question 1 for the months of July and August while she is working as a live-in nanny. At the end of August she returns the TV and goes back to her parents' home to attend school.

**a)** How much did Jasmine pay for the use of the TV for two months?
**b)** Why might the rent-to-own option be reasonable for Jasmine?

**3.** Some rent-to-own stores have **early buy-out** options. Here is one: If the item is bought within the first three months, 100% of rental payments apply to the purchase. If it is bought after the first three months, 50% of rental payments apply to the purchase.

Yung rents the same television model as in question 1 from a store with this early buy-out option. He decides to buy it after two payments. How much more must he pay to buy it?

**4.** Marita rents the same television model from the store with the early buy-out option in question 3. She decides to buy it after eight payments.

**a)** How much has she already paid in rent?
**b)** How much of that payment is applied to the purchase?
**c)** How much more must she pay to buy it?
**d)** How much does she pay in total?

## Practise

5. Kalyn enters into a rent-to-own agreement for a washer and dryer with a cash price of $1043. The maximum rental period is 30 months. The rental fee is $80 per month.
   a) What is the total cash price of the washer and dryer, including taxes?
   b) What does Kalyn pay if she rents the washer and dryer for four months?
   c) What does she pay if she rents the washer and dryer for the entire 30 months? Compare that total to the cash price.
6. Darren rents the same washer and dryer models as in question 5. He decides to buy them after two payments. The same buy-out option as in question 3 applies. How much more must he pay to buy the washer and dryer?
7. Curtis enters into a rent-to-own agreement for a refrigerator with a cash price of $574. The maximum rental period is 24 months. The rental fee is $55 per month.
   a) What is the total cash price of the refrigerator, including taxes?
   b) What does Curtis pay if he rents the refrigerator for eight months?
   c) What does he pay if he rents the refrigerator for 24 months? Compare that total to the cash price.
8. Joanna rents the same refrigerator model as in question 7. She decides to buy it after seven payments. The same buy-out option as in question 3 applies.
   a) How much has she already paid in rent?
   b) How much of what she has paid is applied to the purchase?
   c) How much more must she pay to buy it?
   d) How much does she pay in total?
9. What are the advantages and disadvantages of renting to own?
10. Justify this statement:
    If you rent for the maximum rental period, you pay more than double the cash price.
11. Consider the early buy-out option discussed. If you are going to buy, when is the best time? Explain.

**Skills Check** **Operations with Decimals**

Use your calculator. Round answers to the nearest cent if necessary.

a) $48 × 50 b) $490 ÷ 12 c) 85 × $14
d) $593 + $49.75 e) $1025 ÷ 18 f) $387.20 + $35
g) $212.25 × 25 h) $1398.22 + $49.95 i) $1845.90 ÷ 12

# 5.8 Buying on an Instalment Plan

Instalment plans allow you to get what you want to buy when you want it. A **deferral charge** or **administration fee** is usually charged for the privilege of not paying the total amount at the time of purchase. You pay the fee and the taxes at the time of purchase.

With some plans the selling price of the purchase is divided into equal monthly payments. That amount must be paid when due or interest will be charged.

Other plans add **interest**, a percent of the selling price, to the selling price and divide that amount into equal monthly payments.

## Develop

WS **Each purchase in questions 1 to 3 has a $45 deferral charge and 12 monthly payments. Answer a) and b).**

**a)** What is the total amount to be paid at the time of purchase?
**b)** What is the amount of each monthly payment?

**1.** Jodi purchases a sofa bed that sells for $999 plus taxes.

**2.** Ted buys a jacket that sells for $219 plus taxes.

**3.** Omar purchases a bedroom suite that sells for $1839 plus taxes.

**4.** A store usually has a fixed deferral charge for a one-year instalment plan. Consider the $45 charge in questions 1 to 3. Who got the best deal—Jodi, Ted, or Omar? Explain.

## Practise

WS **For questions 5 to 7, each purchase has a $39 deferral charge and 20 monthly payments. Answer a) to d).**

**a)** What is the cash price, including taxes?
**b)** What is the total amount to be paid at the time of purchase?
**c)** What is the total cost of paying by instalments?
**d)** How much more does it cost to pay by instalments?

**5.** Jacob pays $152.70 a month for a $2588 plus taxes dining room suite.

**6.** Fran pays $23.55 a month for a $399 plus taxes kitchen table and chairs.

**7.** Lynne pays $64.85 a month for a $1099 plus taxes sofa.

# 5.9 No Interest or Payments for a Specified Time

Plans that charge no interest or demand no payments for 6 months, a year, or two years may sound very inviting. Most stores that offer such plans have their own credit cards. If you don't pay before the specified time is up, they charge you interest from the time of purchase.

Just as with instalment plans, a buyer pays a **deferral charge** or **administration fee**, as well as the taxes, at the time of purchase. The remaining amount, the original selling price, is to be paid before the specified time is up.

## Explore

Why might paying an instalment each month be better for some people than paying the selling price just before the specified time is up?

## Develop

WS **For questions 1 to 3, each purchase is made on a no interest or payments for one year basis. The administration fee is $49.95. Answer the following.**

**a)** What is the total amount to be paid at the time of purchase?
**b)** What is the remaining amount to be paid before one year is up?
**c)** What is the total cost if the items are paid for before one year is up?
**d)** What amount of interest is charged if the payment is not made before one year is up?

**1.** Juanita purchases a $999 plus taxes computer. She will get a bill for $1299 if she has not paid the original selling price before one year is up.

**2.** Barbara purchases a $579 plus taxes TV. She will get a bill for $780.02 if she has not paid the original selling price before one year is up.

**3.** Donald purchases a $799 plus taxes micro-component system. He will get a bill for $1075.90 if he has not paid the original selling price before one year is up.

**4. a)** List the advantages and disadvantages of each of these options:
- layaway
- renting with an option to buy
- instalments
- no interest or payments for a specified time

**b)** Which option, if any, would you prefer to use for a major purchase, such as electronics, furniture, or appliances? Why?

# 5.10 Putting It All Together: Deciding What to Buy and How to Pay

WS **1. a)** Use newspaper ads, flyers, catalogues, and/or the Internet to select at least three camcorder models.

**b)** List up to six features for each model, including price.

**c)** Select one model to buy, and justify your decision.

**2.** Assume a camcorder that you liked was priced $419.99 U.S.

**a)** What is its price in Canadian dollars using the current conversion rate?

**b)** How much PST and GST must be paid?

**c)** What other costs might you have to pay before receiving your purchase?

Access to Web sites selling camcorders can be gained through the *Mathematics for Everyday Life 11* page of irwinpublishing.com/students.

**3.** Determine the total cost, including taxes, of the model you selected in question 1 c). Did you have to convert the price to Canadian dollars?

**4.** Assume you buy the camcorder you selected in question 1 from a store where you earn 50 points for every before-taxes dollar spent. How many points would you earn?

**5.** Determine the total cost of the camcorder you selected in question 1 under each of these options.

**a)** layaway
- pay 10% down
- pay the remaining amount and get the camcorder in 4 months

**b)** instalments
- pay a $45 deferral charge and the taxes at the time of purchase and get the camcorder
- pay the original selling price divided into 12 equal monthly payments

**c)** no interest or payments for a year
- pay a $49.95 administration fee and the taxes at the time of purchase and get the camcorder
- pay the original selling price before one year is up

**6. a)** Which payment option from question 5 would you prefer to use? Why?

**b)** Given your income and savings from "your job" in Section 2.8, can you afford the camcorder using that payment option? Explain.

## 5.11 Chapter Review

**1.** A brand of cheese is sold in 340 g packages for $3.79 and 600 g packages for $5.29.

**a)** Calculate the unit price per gram for each package size.
**b)** Calculate the unit price per 100 g for each package size.
**c)** Which package has the lower unit price?
**d)** What other considerations should be taken into account before selecting the package size?

**2.** Swimming passes are sold for $2.75 each, five for $11.50, ten for $21, and 25 for $45. Madeline intends to swim ten times to prepare for a mini-triathlon.

**a)** How much would it cost to buy ten individual passes?
**b)** How much would it cost to buy two 5 packs?
**c)** How much would be saved by buying a 10 pack over ten individual passes?
**d)** What is the unit price per swim for each of 5, 10, and 25 packs?
**e)** Why is it not a better buy for Madeline to purchase a 25 pack?

**3.** Describe two different incentives that stores use to encourage shopping at their stores.

**4.** Estimate and calculate prices in Canadian dollars. Use this conversion factor: $1.00 U.S. = $1.545 00 Cdn.

**a)** $24.99 U.S. **b)** $105.39 U.S. **c)** $52.89 U.S.

**5.** Adam was shopping in an outlet mall in Detroit on an overnight trip. He bought a camera that cost $219 U.S., when $1.00 U.S. = $1.545 00 Cdn.

**a)** Estimate the cost in Canadian dollars.
**b)** Calculate the cost in Canadian dollars.
**c)** How much does Adam pay in GST and PST?

**6.** Assume that you plan to purchase each of the following items. List at least four features of each and rank them in order of importance to you.

**a)** winter boots **b)** stereo system **c)** barbecue

**7.** A washing machine sells for $699 plus taxes. Ernie buys one for his family and pays a $45 deferral charge and the taxes at the time of the purchase. He makes 20 monthly payments of $41.25.

**a)** What is the cash price, including taxes?
**b)** What is the total amount to be paid at the time of purchase?
**c)** What is the total cost of paying by instalments?
**d)** How much more does it cost to pay by instalments?

**8.** A refrigerator selling for $829 plus taxes can be purchased with no interest or payments for a year. An administration fee of $48.95 and the taxes are paid at the time of purchase. If the remaining $829 is not paid before the year is up, a bill for $1105.80 is sent.

**a)** What is the total amount to be paid at the time of purchase?

**b)** What is the total cost if the refrigerator is paid for before one year is up?

**c)** What amount of interest is charged if the payment is not made before one year is up?

**9.** Sandra is starting a business making custom window coverings. She wants to buy a computerized sewing machine that sells for $1699 plus taxes. She is considering options to pay for the machine.

*Option 1 Layaway*
10% down; payment of remainder on pickup in up to 6 months

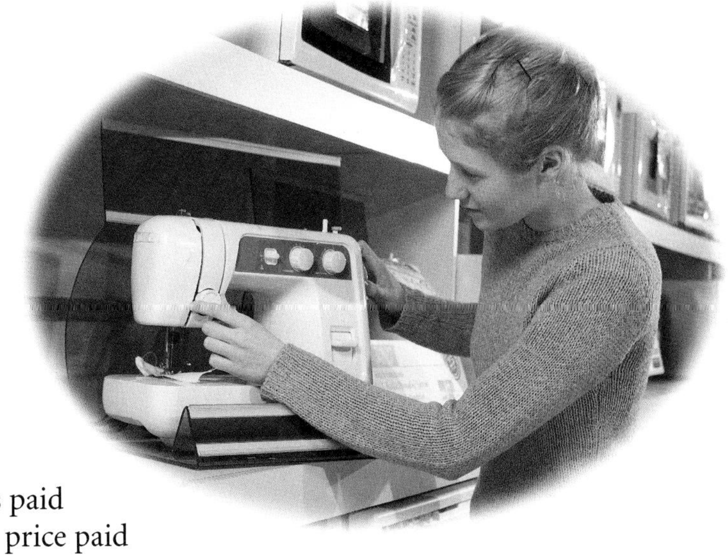

*Option 2 Rent-to-Own*
monthly rental fee of $162.92; maximum rental period of 24 months

*Option 3 Instalment Plan A*
deferral charge of $39 and the taxes at the time of purhcase; 20 monthly payments of $100.25

*Option 4 Instalment Plan B*
deferral fee of $34.95 and the taxes paid at the time of purchase; the selling price paid in 12 equal monthly payments

*Option 5 No Interest or Payments for 12 Months*
administration fee of $44.95 and the taxes paid at the time of purchase; the selling price paid before 12 months are up

**a)** How much more than the total cash price including taxes would Sandra pay for the machine under each option?

**b)** Which option would you recommend to Sandra and why?

CHAPTER

# 6 Banking Transactions and Saving Money

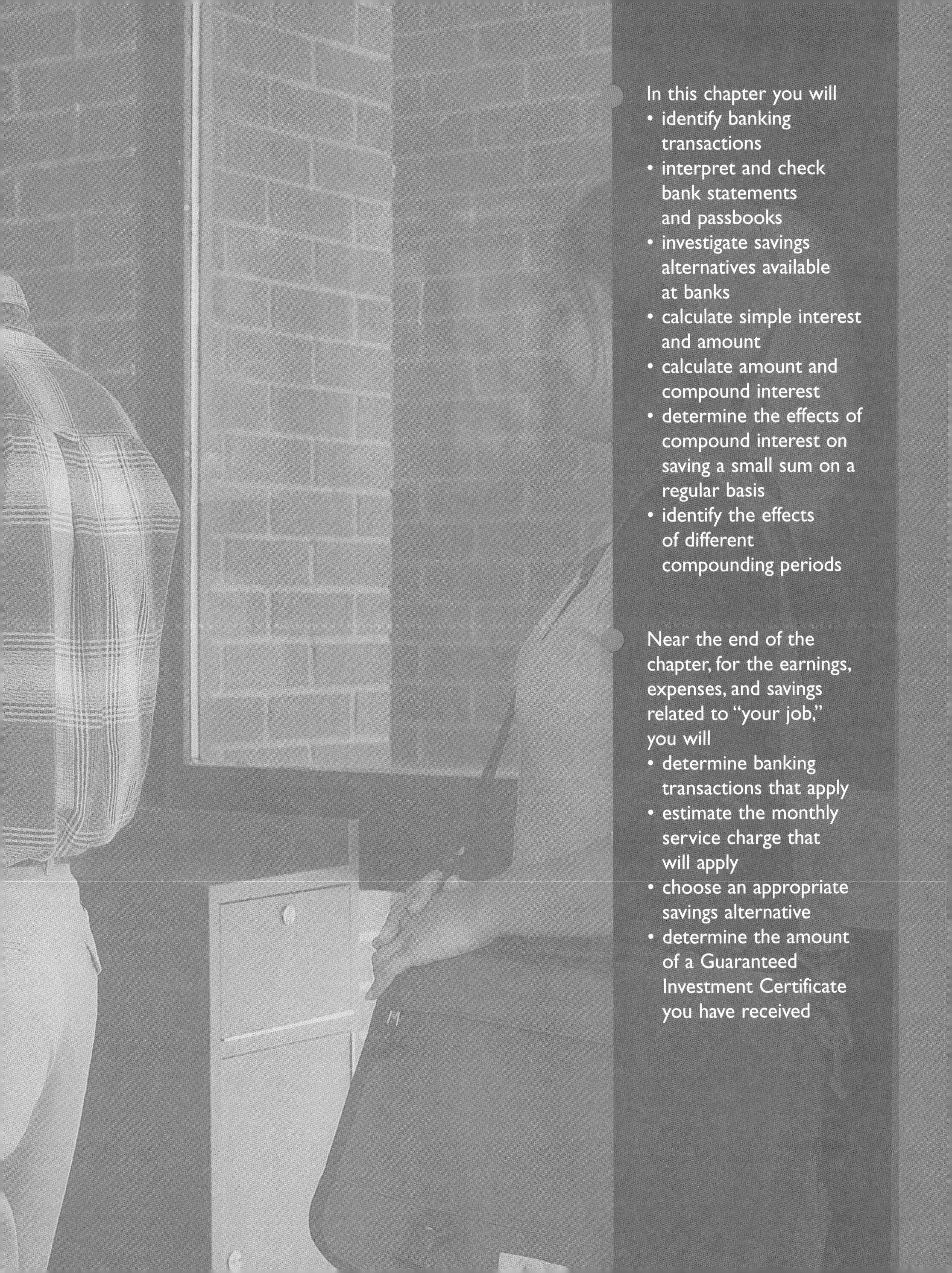

In this chapter you will

- identify banking transactions
- interpret and check bank statements and passbooks
- investigate savings alternatives available at banks
- calculate simple interest and amount
- calculate amount and compound interest
- determine the effects of compound interest on saving a small sum on a regular basis
- identify the effects of different compounding periods

Near the end of the chapter, for the earnings, expenses, and savings related to "your job," you will

- determine banking transactions that apply
- estimate the monthly service charge that will apply
- choose an appropriate savings alternative
- determine the amount of a Guaranteed Investment Certificate you have received

# 6.1 Banking Transactions

Banks and other financial institutions provide a safe place to keep your money and a means to pay expenses.

When you deposit money at a bank or other financial institution, it may pay you **interest**, a percent of the amount deposited, for keeping your money there. It also charges fees for **transactions**, such as depositing, transferring, and withdrawing money, writing cheques, and paying bills. Unless you have a sizable sum of money, you pay the bank more in fees than the bank pays you in interest.

Banks offer a variety of **chequing accounts**. Most people use this type of account for the transactions associated with depositing income and paying expenses. These accounts pay very little interest or no interest, and they have monthly fees and/or transaction fees.

## Explore

What are automated teller machines (ATMs) used for? Where are they found? What do you need to use them? Does it cost to use them?

## Develop

1. Banks encourage us to use **self-service** or **automated banking** rather than **full-service banking** at tellers. **Automated teller machines (ATMs)** are one type of self-service banking. The following transactions, which can be made at a teller, can be made at ATMs associated with your bank.
   - withdrawing money
   - depositing money (cheques and cash)
   - paying bills
   - transferring money between accounts
   - checking the balance in your account

   Services such as opening an account and buying foreign money are available only through full-service banking.

   At ATMs associated with other banks or non-bank ATMs, you can do only one of these transactions. Which is it? Why would that be?

2. ATMs are also called ABMs. What do you think "ABM" stands for?

3. **Writing a cheque** is considered self-service banking. Why would it be self-service rather than full-service?

4. **Telephone banking** and **on-line banking** are two other types of automated or self-service banking. Which transactions listed in question 1 would you not be able to do by telephone or on-line? Why?

5. **Electronic funds transfer** is another type of automated banking.
   - **Direct deposits** of income from employers, from governments, and from investments can be made to your account.
   - As well as paying bills by the self-service and full-service banking methods already discussed, **automatic withdrawals** from your account can be made for payments on their due dates.
   - Many stores and businesses allow you to pay electronically using your bank card as a **debit card**. Funds are electronically transferred from your account to theirs.

   **a)** List all the automated or self-service banking transactions from questions 1 to 5. You should have 15 transactions.

   **b)** Which, if any, have you used?

6. The fees you pay for transactions depend upon the type of account you have. Usually, you pay a monthly fee that entitles you to a certain number of transactions. For each transaction over that number, you pay a transaction fee.

   Use brochures or the Internet to investigate one type of chequing account offered by one bank and answer the following.

   **a)** What is the name of the bank and the name of the account?

   **b)** How much is the monthly fee?

   **c)** How many full-service (at a teller) transactions do you get for the monthly fee?

   **d)** How many self-service (ATM, cheque writing, telephone, on-line, electronic transfer) transactions do you get for the monthly fee?

   **e)** How much does each additional full-service transaction cost?

   **f)** How much does each additional self-service transaction cost?

Access to Web sites about banks can be gained through the *Mathematics for Everyday Life 11* page of irwinpublishing.com/students.

## Practise

**For questions 7 to 10, use the following chequing account information.**

For a monthly fee of \$4.15, you get
- 8 self-serve transactions
- 5 full-serve transactions

Additional self-serve transactions cost 50¢ each.

Additional full-serve transactions cost \$1 each.

Each use of an ATM not associated with this bank costs \$1.25 plus the transaction fee, if applicable.

**7. a)** The first banking Loraine did one month was to deposit a cheque, withdraw some cash, and pay a bill at an ATM at her bank. How much did that trip to the ATM cost her in fees?

**b)** Later in the month, when Loraine had made more than 8 self-serve transactions, she withdrew cash from an ATM not associated with her bank. The next day she deposited a cheque and paid a bill at an ATM at her bank. How much did those trips to ATMs cost her in fees?

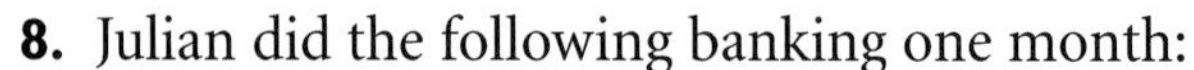

**8.** Julian did the following banking one month:
- deposited a cheque at a teller 2 times
- wrote 3 cheques
- paid 6 bills on-line
- used his debit card 6 times
- withdrew money 3 times from an ATM not associated with his bank

What was the total monthly charge?

**9.** Kelly did the following banking one month:
- deposited a cheque at a teller 2 times
- withdrew money 2 times at an ATM not associated with her bank
- used her debit card 4 times

What was the total monthly charge?

**10.** Sherwin made the following transactions one month:
- deposited a cheque at an ATM at his bank 2 times
- withdrew money 4 times at an ATM at his bank
- paid 9 bills on-line
- used his debit card 8 times

What was the total monthly charge?

**11.** What would the total monthly charge be for each person if they had the chequing account that you researched in question 6?

**a)** Julian in question 8 **b)** Kelly in question 9 **c)** Sherwin in question 10

# 6.2 Career Focus: Security Guard

Steven is a security guard. His job involves

- driving an armoured truck
- delivering cash to non-bank ATMs in malls and stores
- keeping records of how much is delivered to each site
- guarding the armoured truck and the cash

When Steven was hired, he received training from his employer about the specifics of his job. He also became **bonded**, which is like being insured against taking money. Bonding requires a police check. Steven could not have been bonded if he had a criminal record or had ever declared bankruptcy (declared legally unable to pay money owed to others).

Driving the armoured truck requires a Type G driver's licence. Steven carries a firearm, for which he also holds a licence.

**1. a)** What does "bonded" mean?

**b)** What types of licences does Steven need for his job?

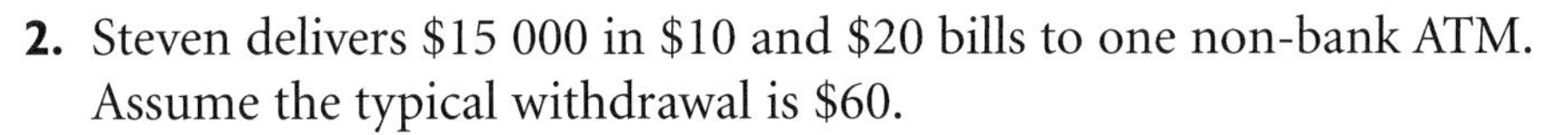

**2.** Steven delivers \$15 000 in \$10 and \$20 bills to one non-bank ATM. Assume the typical withdrawal is \$60.

**a)** How many ATM user fees would be charged by the time \$15 000 is withdrawn?

**b)** At \$1.25 per use, how much is charged in user fees?

**3.** Part of driving the armoured truck is planning the route and scheduling the time for the deliveries. First, Steven picks up the truck from his employer's depot. Next, he picks up the cash for the deliveries. Finally, he delivers the cash and stocks each machine.

Assume your home is a security truck depot and your school is where the cash for the deliveries is picked up. Plan a route and a schedule for the delivery of cash and stocking of several ATMs in your area.

# 6.3 Bank Statements and Passbooks

## Explore

Do you have a bank account? If so, do you have a passbook or do you receive a monthly statement? Where does a passbook or statement tell you what codes such as CHQ, DBT, and WD mean?

## Develop

**Use these codes to answer questions 1 to 3.**

| | | | |
|---|---|---|---|
| ATM | automated teller machine | ET | electronic funds transfer |
| CHQ | cheque | SC | service charge |
| DBT | debit card | WD | withdrawal |
| DEP | deposit | | |

**1.** Some bank accounts offer passbooks. Passbooks vary from bank to bank. It is up to you to keep your passbook up-to-date. Here is the part of Jalinda's passbook that shows her transactions in May.

| DATE | ACCESS POINT | WITHDRAWAL DESCRIPTION | DEPOSIT DESCRIPTION | BALANCE |
|---|---|---|---|---|
| MAY 05 | ET | | PAY:SUNMORE 653.95 | 1284.94 |
| MAY 10 | ATM | WD 50.00 | | 1234.94 |
| MAY 11 | DBT | WD 27.50 | | 1207.44 |
| MAY 19 | ET | | PAY:SUNMORE 653.95 | 1861.39 |
| MAY 25 | | CHQ#234 730.00 | | 1131.39 |
| MAY 28 | ATM | WD 250.00 | | 881.39 |
| MAY 31 | | SC 6.65 | | 874.74 |

**a)** When did Jalinda get paid by her employer, SunMore? Did she deposit her cheque or was it automatically deposited?
**b)** When was the **service charge**, which consists of the monthly and transaction fees, withdrawn?
**c)** What balance did she end the month with?
**d)** What transaction did she make May 11?
**e)** What transaction did she make May 28?
**f)** Explain how the amounts in the Balance column are calculated.
**g)** Why do you think there is no Access Point information given for the transactions on May 25 and May 31?

2. Here is the part of Lena's passbook that shows her transactions in September.

| DATE | ACCESS POINT | WITHDRAWAL DESCRIPTION | | DEPOSIT DESCRIPTION | | BALANCE |
|---|---|---|---|---|---|---|
| SEPT 08 | | CHQ#286 | 23.00 | | | 241.86 |
| SEPT 12 | DBT | WD | 21.60 | | | 220.26 |
| SEPT 19 | | | | CASH | 45.60 | 265.86 |
| SEPT 23 | ET | | | PAY:FRYS PLUS | 230.00 | 495.86 |
| SEPT 25 | ATM | WD | 50.00 | | | 445.86 |
| SEPT 30 | | SC | 5.40 | | | 440.46 |

**a)** When did Lena get paid by her employer, Frys Plus?
**b)** What transaction did she make Sept. 12?
**c)** What transaction did she make Sept. 25?
**d)** How much was the monthly service charge on her account?
**e)** What was the balance in her account on Sept. 23?
**f)** What was the balance in her account on Sept. 24?
**g)** How must she have deposited the cash on Sept. 19? How do you know?

3. Craig doesn't have a passbook for his chequing account. His bank sends him a monthly statement. Here is his statement for October.

| DESCRIPTION | WITHDRAWALS | DEPOSITS | DATE | BALANCE |
|---|---|---|---|---|
| BALANCE FORWARD | | | OCT 01 | 462.98 |
| CHQ#293 | 150.00 | | OCT 05 | 312.98 |
| DBT WD | 76.90 | | OCT 17 | 236.08 |
| ATM WD | 20.00 | | OCT 19 | 216.08 |
| PAY:MILAN'S | | 245.00 | OCT 25 | 461.08 |
| ATM WD | 25.00 | | OCT 25 | 270.00 |
| ATM WD | 15.00 | | OCT 26 | 255.00 |
| ATM WD | 10.00 | | OCT 27 | 245.00 |
| ATM WD | 20.00 | | OCT 28 | 225.00 |
| ATM WD | 5.00 | | OCT 30 | 220.00 |
| SC | 11.65 | | OCT 31 | 208.35 |

**a)** What was Craig's balance at the beginning of the month?
**b)** What was his balance at the end of the month?
**c)** What transaction did he make on Oct. 5?
**d)** What transactions did he make on Oct. 25?
**e)** How much was the monthly service charge on his account?
**f)** Apply what you learned in Section 6.1 and in this section to advise Craig on how to reduce his service charge.

4. Which method of keeping track of transactions do you prefer—a passbook or a monthly statement? Why?

# 6.4 Types of Savings

## Explore

Suppose you have more money coming in than you have going out in essential expense payments each month. What would you do with the surplus money? How could you use the surplus money to earn money?

## Develop

1. One way to earn money at a bank is by depositing money in a **savings account**. Banks offer a variety of savings accounts which usually do not charge a monthly fee. However, it is common to pay transaction fees. Savings accounts pay more interest than chequing accounts. Like chequing accounts, they give you ready access to your money.

   Use brochures or the Internet to investigate one type of savings account offered by one bank and answer the following.

   **a)** What is the name of the bank and the name of the account?
   **b)** What fees are involved?
   **c)** What interest rate is paid?
   **d)** Identify five new terms.
   **e)** Did you read about "interest compounding"? Compound interest is the most common type of interest paid by banks. You will study it later in this chapter.

Access to Web sites about banks can be gained through the *Mathematics for Everyday Life 11* page of irwinpublishing.com/students.

2. Another way to earn money at a bank is by buying a **Guaranteed Investment Certificate (GIC)**. You buy it for a set period of time, called a **term**, and at the end it comes due, or matures. You do not have access to the money until the GIC matures. You are paid more interest than a savings account pays. The interest rate varies depending upon the term.

   **a)** Which do you think would pay a higher rate of interest—a short-term GIC or a long-term GIC? Why?
   **b)** Use brochures or the Internet to investigate one bank's GICs to determine common terms and to verify your answer to part a).

3. A third way to earn money at a bank is by buying **mutual funds**. Mutual funds are collections of stocks that trade on the stock market and are owned by groups of people. It is possible to earn more money buying mutual funds, and stocks in general, than buying GICs. However, you also risk losing money. Although you can sell mutual funds at any time, buying mutual funds, or any stocks, is not recommended for short-term savings.

   Use brochures or the Internet to investigate one bank's mutual funds and any fees associated with buying or selling them.

4. Mutual funds are sold by the unit. This graph shows the performance of one mutual fund.

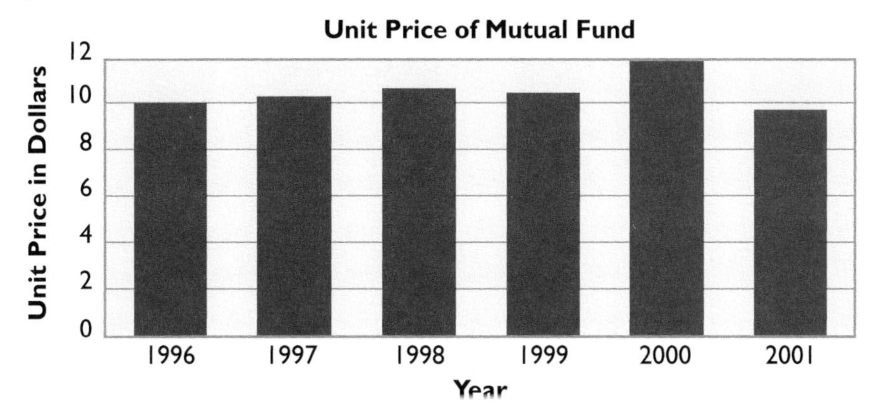


   Assume you bought this mutual fund in 1996.

   **a)** Which year would have been the best for selling your fund? Explain.

   **b)** Which year would have been the worst for selling your fund? Explain.

5. List the advantages and disadvantages of each way to earn money.

   **a)** savings account **b)** GIC **c)** mutual funds

6. If you had $2000 to put towards savings, which savings option offered by a bank would you use in each situation given? Justify your answers.

   **a)** You want to take a trip in four years.

   **b)** You want a down payment for a car in three months.

   **c)** You are saving for your retirement.

## Skills Check — Operations with Decimals

Calculate. Round answer to two decimal places, if necessary.

**a)** $475 \times 0.07 \times 2.5$ **b)** $3850 \times 0.08 \times 3.5$ **c)** $14\,000 \times 0.075 \times 9$

**d)** $(12\,500)(0.05)(6)$ **e)** $(550)(0.045)(4.5)$ **f)** $(850)(0.12)(5.5)$

**g)** $(1250)(0.06)\left(\frac{21}{12}\right)$ **h)** $(2000)(0.07)\left(\frac{21}{12}\right)$ **i)** $(975)(0.03)\left(\frac{15}{12}\right)$

# 6.5 Simple Interest

Compound interest is much more common than simple interest. However, to understand compound interest, you need to understand **simple interest**.

You deposit \$1000 and a year later you have \$1050. The increase in value, \$50, is the interest. The money you deposited, \$1000, is called the **principal**. The money you end up with, \$1050, is called the **amount**.

## Explore

Josh earns \$60 in interest. Mina earns \$70 interest. What are some possible reasons Mina earns more?

## Develop

Simple interest is calculated using this formula.

$I = Prt$ where $I$ is the interest, money earned, in dollars
$P$ is the principal, money deposited, in dollars
$r$ is the rate given as a percent
$t$ is the time, in years

The amount is calculated using this formula.

$A = P + I$ where $A$ is the amount after interest, in dollars
$P$ is the principal, in dollars, as above
$I$ is the interest, in dollars, as above

### Example 1

Laurie deposits \$5000 at 6% **per annum** (per year) for 3 years.

**a)** How much simple interest does she earn altogether?

**b)** How much simple interest does she earn per year?

**c)** What is the amount of the investment after 3 years?

### Solution

**a)** Substitute $P = \$5000$
$r = 6\%$ or $0.06$
$t = 3$ years

into $I = Prt$

$I = (5000)(0.06)(3)$

$I = 900$

> Remember (5000)(0.06)(3) means 5000 × 0.06 × 3.

Laurie earned \$900 simple interest.

b) interest per year = total interest ÷ number of years

$$= 900 \div 3$$
$$= 300$$

Laurie earned $300 simple interest per year.

c) Substitute $P = \$5000$

$I = \$900$

into $A = P + I$

$$A = 5000 + 900$$
$$A = 5900$$

The amount after 3 years is $5900.

## Example 2

Cesar deposits $800 at 5% per annum (per year) for 9 months.

**a)** How much simple interest does he earn?

**b)** What is the amount after 9 months?

### Solution

**a)** 9 months $= \frac{9}{12}$ year

Substitute $P = \$800$

$r = 5\%$ or $0.05$

$t = \frac{9}{12}$ year

into $I = Prt$

$$I = (800)(0.05)\left(\frac{9}{12}\right)$$
$$I = 30$$

He earned $30 simple interest.

**b)** Substitute $P = \$800$

$I = \$30$

into $A = P + I$

$$A = 800 + 30$$
$$A = 830$$

The amount after 9 months is $830.

## Practise

**1.** Calculate the interest earned on each deposit.

**a)** $9000 deposited at 5% per annum for 3 years

**b)** $5000 deposited at 4% per annum for 2 years

**c)** $1000 deposited at 6% per annum for 15 months

**2.** Kim invests $4500 in a 3-year GIC. The interest rate is 5% per annum.

**a)** How much interest does he earn altogether?

**b)** How much interest does he earn per year?

**c)** What is the amount after 3 years?

**3.** Claire invests $9200 in an 8-month GIC. The interest rate is 8% per annum.

**a)** How much interest does she earn?

**b)** What is the amount after 8 months?

4. Sydney invests $1500 in a 21-month GIC. The interest rate is 4% per annum.
   a) How much interest does he earn?
   b) What is the amount after 21 months?

5. Connie saved money for a trip she planned to take in January 2002.
   a) On Jan. 4, 1999, she deposited $1500 at 5% per annum for 3 years. Find the interest earned.
   b) On Jan. 4, 2000, she deposited $2000 at 4% per annum for 2 years. Find the interest earned.
   c) On Jan. 4, 2001, she deposited $1800 at 6% per annum for 1 year. Find the interest earned.
   d) How much money did Connie have on Jan. 4, 2002, for her trip?

6. a) Predict which will be worth more when it comes due:
   $4000 invested at 6% per annum for 5 years
   or $4000 invested at 4% per annum for 7 years.
   b) Determine each amount to check your prediction.

7. a) Predict which will be worth more when it comes due:
   $2000 invested at 5% per annum for 2 years
   or $4000 invested at 5% per annum for 1 year.
   b) Determine each amount to check your prediction.

8. a) When you deposit money in a savings account or buy a GIC, why do you think the bank pays you interest on your money?
   b) Why would GICs pay a higher interest rate than a savings account?

9. Every fall, banks have **Canada Savings Bonds** for sale. Use brochures or the Internet to investigate Canada Savings Bonds. Identify four features of Canada Savings Bonds. Which features are like GICs? Which are different?

Access to the Canada Savings Bonds Web site can be gained through the *Mathematics for Everyday Life 11* page of irwinpublishing.com/students.

**Skills Check** **Mental Math**

Use mental math to calculate.

a) 8 transactions at 50¢ each
b) 12 transactions at 80¢ each
c) 4 uses of an ATM at $1.25 each
d) 5 transactions at 75¢ each
e) 9 transactions at 60¢ each
f) 7 uses of an ATM at $1.25 each

# 6.6 From Simple Interest to Compound Interest

## Explore

Which option do you think would be worth more? Why?

- $1000 invested at 7% per annum for 2 years
- $1000 invested at 7% per annum for 1 year and then the $1000 plus the interest after 1 year invested for a second year

## Develop

**1.** For $1000 invested at 5% simple interest per annum for 20 years, calculate the following.

**a)** the interest earned altogether
**b)** the interest earned per year
**c)** the amount after 20 years

**2.** Use this partial spreadsheet.

| | A | B | C | D |
|---|---|---|---|---|
| 1 | Principal ($): | | 1000.00 | |
| 2 | Interest Rate (%): | | 5 | |
| 3 | | | | |
| 4 | Year | Principal ($) | Interest ($) | Amount ($) |
| 5 | 1 | 1000.00 | 50.00 | 1050.00 |
| 6 | 2 | 1050.00 | 52.00 | 1102.50 |
| 7 | 3 | 1102.50 | 55.13 | 1157.63 |
| 8 | 4 | 1157.63 | 57.88 | 1215.51 |

**a)** Describe how the values in C5 and D5 were calculated.
**b)** Which value in Row 5 does the value in cell B6 equal?

AT **3.** Examine the full results of the spreadsheet for the $1000 investment.

**a)** What do you notice about the interest as the years increase?
**b)** What is the amount after 20 years?
**c)** How much interest is earned altogether?
**d)** After how many years is the amount double the initial investment?

AT **4.** Examine the simple interest calculations in question 1 and the spreadsheet from question 3.

**a)** Compare the yearly interest earned.

**b)** Compare the interest earned altogether. Which is greater and by how much?

**c)** Compare the amount after 20 years. Which is greater and by how much?

**d)** After how many years is that amount in the spreadsheet close to the amount for simple interest?

AT **5.** **a)** Select cells D4 to D24 of the spreadsheet from question 2. Create a line graph. Title the graph (chart) "Amount of $1000 at 5%." Title the X-axis "Year" and the Y-axis "Amount ($)."

**b)** Does the graph show a straight line or a curve?

**c)** What does the shape of the graph indicate?

**6.** For $5000 invested at 6% simple interest per annum for 20 years, calculate the following.

**a)** the interest earned altogether

**b)** the interest earned per year

**c)** the amount after 20 years

AT **7.** Use the spreadsheet from question 3.

**a)** Change the principal to $5000 by entering 5000 in cell C1.

**b)** Change the rate to 6% by entering 6 in cell C2.

**c)** What do you notice about the interest as the years increase?

**d)** What is the amount after 20 years?

**e)** How much interest is earned altogether?

**f)** After how many years is the amount double the initial investment?

AT **8.** Examine the simple interest calculations in question 6 and the results of the spreadsheet in question 7.

**a)** Compare the yearly interest earned.

**b)** Compare the interest earned altogether. Which is greater and by how much?

**c)** Compare the amount after 20 years. Which is greater and by how much?

**d)** After how many years is that amount in the spreadsheet close to the amount for simple interest?

AT **9.** **a)** Select cells D4 to D24 of the spreadsheet from question 6. Create a line graph. Title the graph (chart) "Amount of $5000 at 6%." Title the X-axis "Year" and the Y-axis "Amount ($)."

**b)** Describe the shape of the graph and what it indicates.

**10.** For $2000 invested at 7% simple interest per annum for 20 years, calculate the following.

**a)** the interest earned altogether
**b)** the interest earned per year
**c)** the amount after 20 years

AT **11.** Use your spreadsheet from question 7.

**a)** Change the principal to $2000 by entering 2000 in cell C1.
**b)** Change the rate to 7% by entering 7 in cell C2.
**c)** What do you notice about the interest as the years increase?
**d)** What is the amount after 20 years?
**e)** How much interest is earned altogether?
**f)** After how many years is the amount double the initial investment?

AT **12.** Examine the simple interest calculations in question 10 and the results of the spreadsheet in question 11.

**a)** Compare the yearly interest earned.
**b)** Compare the interest earned altogether. Which is greater and by how much?
**c)** Compare the amount after 20 years. Which is greater and by how much?
**d)** After how many years is that amount in the spreadsheet close to the amount for simple interest?

AT **13.** Select Column D in your spreadsheet from question 10. Predict the shape when the amounts are plotted in a line graph. Create the graph. Title the graph (chart) "Amount of $2000 at 7%." Title the X-axis "Year" and the Y-axis "Amount ($)."

**14.** Using the spreadsheet and the simple interest formula, you calculated the interest and amount for each of the following **compound interest** scenarios.

- $1000 at 5% per annum, compounded annually for 20 years
- $5000 at 6% per annum, compounded annually for 20 years
- $2000 at 7% per annum, compounded annually for 20 years

**a)** Why do you think interest such as that is called compound interest?
**b)** Why is a spreadsheet useful for calculating compound interest using the simple interest formula?
**c)** How is compound interest different from simple interest?

# 6.7 Watching Savings Grow

## Explore

Imagine investing $100 each year on your birthday. What factors would affect the amount of money you had saved by the time you turned 40?

## Develop

**1.** Zarina will be 17 on her next birthday. She is going to invest $100 at 6% per annum, compounded annually for 23 years. She plans to invest $100 on each birthday up to and including her 39th birthday.

This spreadsheet can be used to determine the amount when she turns 40 if each year she has saved $100 to invest and the rate stays at 6%.

| | A | B | C | D |
|---|---|---|---|---|
| 1 | Zarina Saved ($): | | 100.00 | |
| 2 | Interest Rate (%): | | 6 | |
| 3 | | | | |
| 4 | Birthday | Principal ($) | Interest ($) | Amount ($) at Next Birthday |
| 5 | 17 | 100.00 | 6.00 | 106.00 |
| 6 | 18 | 206.00 | 12.36 | 218.36 |
| 7 | 19 | 318.36 | 19.10 | 337.46 |

Describe how the values in C5, D5, and B6 were calculated.

AT **2.** Examine the full results of the spreadsheet for Zarina.

**a)** What is the amount when Zarina turns 40?
**b)** What is the amount when she turns 30?
**c)** After which birthday was the interest more than $100?

AT **3.** Select the entries in cells D4 to D27 of your spreadsheet. Create a line graph. Title the graph (chart) "Investing $100 Each Year." Title the X-axis "Number of Years" and the Y-axis "Amount ($)."

AT **4.** Suppose Zarina was able to save $500 each year, but the interest rate was only 5% per annum. Use the spreadsheet from question 2.

**a)** Change the saved values to $500 by entering 500 in cell C1.
**b)** Change the rate to 5% by entering 5 in cell C2.
**c)** What is the amount when Zarina turns 40?
**d)** What is the amount when she turns 35?
**e)** What is the amount when she turns 25?
**f)** After which birthday was the interest more than $500?

AT **5.** Suppose Zarina was able to save $200 each year, but the interest rate was 7% per annum. Use your spreadsheet from question 4.

**a)** Change the saved values to $200 by entering 200 in cell C1.
**b)** Change the rate to 7% by entering 7 in cell C2.
**c)** What is the amount when Zarina turns 40?
**d)** What is the amount when she turns 30?
**e)** What is the amount when she turns 25?
**f)** After which birthday was the interest higher than the amount you had her saving each year?

**6.** Choose how much Zarina is able to save each year and decide on an interest rate between 3% and 12% per annum. Use your spreadsheet from question 5.

**a)** Change the saved values to the amount you chose by entering the number in cell C1.
**b)** Change the rate to the percent you decided on by entering the number in cell C2.
**c)** What is the amount when Zarina turns 40?
**d)** What is the amount when she turns 35?
**e)** What is the amount when she turns 30?
**f)** After which birthday was the interest higher than the amount you had her saving each year?

## Skills Check Operations with Decimals

Calculate. Round answer to two decimal places, if necessary.

**a)** $2.5(12.8 + 33.7)$ **b)** $1.04^5$ **c)** $1 + 6 \div 36\ 500$
**d)** $525(1 + 0.07)$ **e)** $(2.7 + 14.9)^3$ **f)** $(1.2 + 0.06)^{12}$
**g)** $12\ 000(1.07)$ **h)** $4.5 + 32.6 \times 2.4$ **i)** $675(1 + 0.07)^{10}$

# 6.8 Compound Interest

In the previous two sections, you calculated compound interest and amount using the formulas for simple interest. The interest was compounded annually, that is, there was one **compounding period** a year. Interest can also compound

- semi-annually (twice a year)
- quarterly (four times a year)

To determine the amount after money has been earning compound interest directly, you need to first determine the interest rate per compounding period, $i$, and the number of compounding periods, $n$.

$i$ = rate per annum ÷ number of compounding periods per year
$n$ = number of compounding periods per year × number of years

$A = P(1 + i)^n$ where $A$ is the amount, in dollars
$P$ is the principal, in dollars
$i$ is the interest rate per compounding period
$n$ is the number of compounding periods

## Develop

### Example 1

Determine $i$ and $n$ for each.

**a)** 5% per annum compounded annually for 4 years
**b)** 6% per annum compounded semi-annually for 10 years
**c)** 8% per annum compounded quarterly for 2 years

#### Solution

| | Rate per Annum, % | Compounded | Compounding Periods in One Year | Rate per Compounding Period, $i$% | Term, years | Number of Compounding Periods, $n$ |
|---|---|---|---|---|---|---|
| **a)** | 5 | annually | 1 | $5 \div 1 = 5$ | 4 | $1 \times 4 = 4$ |
| **b)** | 6 | semi-annually | 2 | $6 \div 2 = 3$ | 10 | $2 \times 10 = 20$ |
| **c)** | 8 | quarterly | 4 | $8 \div 4 = 2$ | 2 | $4 \times 2 = 8$ |

### Example 2

Tania invests $1000 in a 3-year GIC at 6% per annum. Interest is compounded semi-annually. Calculate the amount of the GIC at the end of 3 years.

## Solution 

Determine the number of compounding periods per year because both $i$ and $n$ require it.
"Compounded semi-annually" means that there are 2 compounding periods per year.

$$\begin{aligned} i &= \text{rate per annum} \div \text{number of compounding periods per year} \\ &= 6 \div 2 \\ &= 3 \end{aligned}$$

3% as a decimal is 0.03.

$$\begin{aligned} n &= \text{number of compounding periods per year} \times \text{number of years} \\ &= 2 \times 3 \\ &= 6 \end{aligned}$$

$P = 1000$

Substitute into
$$\begin{aligned} A &= P(1 + i)^n \\ &= 1000(1 + 0.03)^6 \\ &= 1000(1.03)^6 \\ &\doteq 1194.05 \end{aligned}$$

The amount of Tania's investment is $1194.05.

## Practise

**1.** Complete the table.

| | Rate per Annum, % | Compounded | Compounding Periods in One Year | Rate per Compounding Period, $i$% | Term, years | Number of Compounding Periods, $n$ |
|---|---|---|---|---|---|---|
| **a)** | 5 | annually | | | 6 | |
| **b)** | 6 | semi-annually | | | 4 | |
| **c)** | 8 | semi-annually | | | 2 | |
| **d)** | 7 | annually | | | 8 | |
| **e)** | 6 | quarterly | | | 3 | |

**2.** Determine the amount if $5000 is invested each way in question 1.

**3.** Ashu invests $4000 at 4% per annum, compounded semi-annually for 5 years.

**a)** What was the amount after 5 years?
**b)** How much interest was earned?

**4. a)** Determine the amount of a $10 000 investment at 4% per annum, compounded annually for 8 years.
**b)** How much interest was earned?

# 6.9 Compounding Periods

Interest is compounded not only annually, semi-annually, and quarterly, but also monthly and even daily.

## Explore

Two GICs have the same principal, the same rate per annum, and the same term. One is compounded semi-annually and the other is compounded daily. Which do you think would earn more interest? How could you confirm your thinking?

## Develop

**1.** Recall that $n$ and $i$ depend upon how often the interest is compounded.
$i$ = rate per annum ÷ number of compounding periods per year
$n$ = number of compounding periods per year × number of years
Complete the table.

| | Rate per Annum, % | Compounded | Compounding Periods in One Year | Rate per Compounding Period, $i$% | Term, years | Number of Compounding Periods, $n$ |
|---|---|---|---|---|---|---|
| **a)** | 12 | annually | | | 2 | |
| **b)** | 12 | semi-annually | | | 2 | |
| **c)** | 12 | quarterly | | | 2 | |
| **d)** | 12 | monthly | | | 2 | |
| **e)** | 12 | daily | | * | 2 | |

*For daily, express as a fraction.

**2.** Determine the amount if $1000 is invested each way in question 1. Remember to express $i$% as a decimal. Except for part e) where interest is compounded daily, use $i = \frac{0.12}{365}$.

**3. a)** Which amount from question 2 was the greatest? Is that what you would have expected? Explain.

**b)** What is the effect of increasing the number of compounding periods in a year?

## Develop

**4.** Use this spreadsheet.

| | A | B | C | D | E | F | G | H |
|---|---|---|---|---|---|---|---|---|
| 1 | Principal, P ($) | Rate per Annum, % | Compounded | Compounding Periods in One Year | Rate per Compounding Period, i % | Term, years | Number of Compounding Periods, n | Amount A ($) |
| 2 | 1000 | 12 | annually | 1 | 12 | 2 | 2 | 1254.40 |
| 3 | | | semi-annually | | | | | |
| 4 | | | quarterly | | | | | |
| 5 | | | monthly | | | | | |
| 6 | | | daily | | | | | |

**a)** Given cell C2, describe how cell D2 was determined.
**b)** Given cells B2 and D2, describe how cell E2 was calculated.
**c)** Given cells F2 and D2, describe how cell G2 was calculated.
**d)** Given cells A2, E2, and G2, describe how cell H2 was calculated.

**5.** Use the spreadsheet from question 4.

**a)** Determine the number of compounding periods in one year for each way of compounding interest. Enter numbers in cells D2 to D6.
**b)** FILL DOWN Column A, Rows 2 to 6, to show that the principal is $1000 for each.
**c)** FILL DOWN Column B, Rows 2 to 6, to show that the rate per annum is 12% for each.
**d)** FILL DOWN Column F, Rows 2 to 6, to show that the term is 2 years for each.
**e)** FILL DOWN Column E, Rows 2 to 6, to calculate *i* for each.
**f)** FILL DOWN Column G, Rows 2 to 6, to calculate *n* for each.
**g)** FILL DOWN Column H, Rows 2 to 6, to calculate *A* for each.

**6. a)** Which amount from question 5 was the greatest? Is that what you would have expected? Explain.
**b)** What is the effect of increasing the number of compounding periods in a year?

# 6.10 Putting It All Together: Banking Transactions and Saving

In Section 2.8, you determined for "your job"

- your take-home pay
- your preferred pay frequency
- your expenses, both essential and non-essential
- your preferred method of paying bills
- how you could save at least $75 a month

**Think about "your job" from the previous chapters to answer the following questions.**

**1.** Would you like to have your pay cheque automatically deposited to your chequing account? Why or why not?

**2.** Which method of paying bills would you use to pay each of the following, assuming you have these expenses? Explain your choices.

**a)** rent
**b)** telephone
**c)** cable
**d)** groceries
**e)** credit card statement

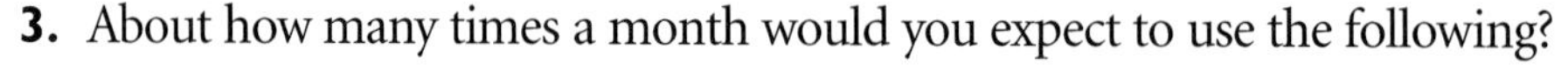

**3.** About how many times a month would you expect to use the following?

**a)** an ATM **b)** your debit card

**4.** Refer to the chequing account that you investigated in Section 6.1. Would you expect the monthly fee to cover all the transactions you would have? Estimate your service charge for a typical month.

**5.** You planned to save at least $75 each month. Which way of saving money offered by banks would you use to save for a camping trip (as in Chapter 4) and to buy a camcorder (as in Chapter 5)? Why?

**6.** Five years ago a family member bought you a $2000 GIC. It has been earning interest at 6% per annum, compounded annually. It is now mature. What is the amount of the GIC?

***In the next chapter you will learn about investment options and what you might do with the money from the GIC.***

# 6.11 Chapter Review

**1.** **a)** Name three banking transactions.
**b)** Name two transactions that can be made at an ATM associated with your bank, but cannot be made on-line or by telephone banking.
**c)** How are banking transactions paid for?

**2.** Here is the part of Ryan's passbook that shows his transactions in May.

| DATE | ACCESS POINT | WITHDRAWAL DESCRIPTION | | DEPOSIT DESCRIPTION | | BALANCE |
|---|---|---|---|---|---|---|
| MAY 09 | ATM | WD | 30.00 | | | 126.98 |
| MAY 17 | DBT | WD | 22.78 | | | 104.20 |
| MAY 21 | ET | | | PAY:CD WORLD | 260.00 | 364.20 |
| MAY 28 | | CHQ#436 | 50.00 | | | 314.20 |
| MAY 31 | | SC | 5.40 | | | 308.80 |

**a)** When did Ryan get paid by his employer, CD World?
**b)** What transaction did he make May 9? May 28?
**c)** Explain how the amounts in the Balance column are calculated.

**3.** You want to make a down payment on a car in three months. You have $3000 now. Would you put the $3000 in a savings account, a GIC, or a mutual fund? Explain your choice.

**4.** Calculate the simple interest and the amount for each.
**a)** $1500 deposited at 5% per annum for 3 years
**b)** $4000 deposited at 6% per annum for 20 months

**5.** Describe the difference between simple and compound interest.

WS **6.** Complete the table.

| | Rate per annum, % | Compounded | Compounding Periods in One Year | Rate per Compounding Period, $i$ % | Term, years | Number of Compounding Periods, $n$ |
|---|---|---|---|---|---|---|
| **a)** | 5 | annually | | | 3 | |
| **b)** | 6 | semi-annually | | | 2 | |
| **c)** | 8 | quarterly | | | 4 | |

**7.** Edwina invests $7000 at 6% per annum, compounded annually for 3 years. What is the amount after 3 years and the interest earned?

A **8.** Show the effect of increasing the number of compounding periods in a year.

CHAPTER

# 7 Investing Money

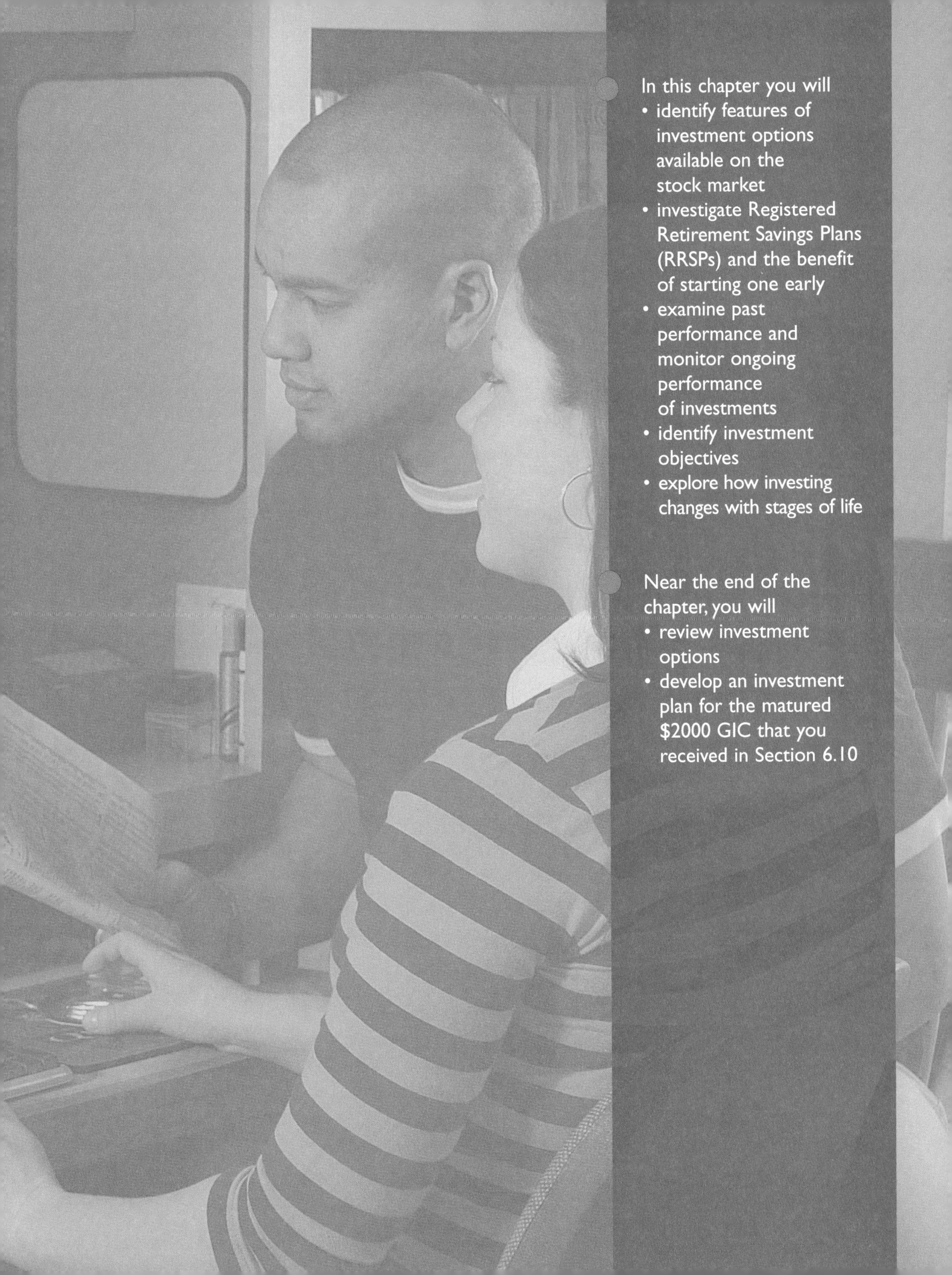

In this chapter you will

- identify features of investment options available on the stock market
- investigate Registered Retirement Savings Plans (RRSPs) and the benefit of starting one early
- examine past performance and monitor ongoing performance of investments
- identify investment objectives
- explore how investing changes with stages of life

Near the end of the chapter, you will

- review investment options
- develop an investment plan for the matured $2000 GIC that you received in Section 6.10

# 7.1 Types of Investments

In Chapter 6, you learned about using surplus money to earn money at a bank. You save money to buy things or take a vacation. Saving is for the short term. When you set money aside and put it to work in order to get more money back in the future, you are **investing**. Investing is for the long term.

## Explore

Use the business section of a national newspaper to identify ways to invest money.

## Develop

**1.** Governments and corporations borrow large sums of money by selling **bonds**. Like GICs and Canada Savings Bonds, bonds pay interest. Unlike GICs and Canada Savings Bonds, they are bought and sold on the stock market by brokers. The **market value**, or price paid for a bond, depends on

- the interest rate that the bond pays
- its maturity date
- the security rating of the government or corporation

As a result, you might buy a bond for more than or less than its **face value**, or stated value, and you could sell it later for more than or less than what you paid.

**a)** How do people earn money with bonds?
**b)** When do people lose money with bonds?

### Example 1

On June 2, 1995, James bought $10 000 worth of General Appliance Corporation bonds that paid 6% simple interest per annum. The bonds mature in 2015. If James sells his bonds before June 2, 2015, he might sell them for more than or less than $10 000.

**a)** Use the simple interest formula, $I = Prt$, to determine how much interest James was paid on June 2, 1996.

**b)** If James holds the bonds to 2015, how much interest will he earn?

**c)** If he sells them for $9200 after owning them for 8 years, how much will his $10 000 have earned in the 8 years?

## Solution 

**a)** $P = \$10\ 000$, $r = 6\%$ or $0.06$, $t = 1$ year

$$I = Prt$$
$$= 10\ 000 \times 0.06 \times 1$$
$$= 600$$

James was paid \$600 simple interest.

**b)** From 1995 to 2015 is 20 years.

$$I \text{ for 20 years} = I \text{ for 1 year} \times 20$$
$$= 600 \times 20$$
$$= 12\ 000$$

James will earn \$12 000 simple interest.

**c)**
$$I \text{ for 8 years} = I \text{ for 1 year} \times 8$$
$$= 600 \times 8$$
$$= 4800$$

$$\text{earnings} = \text{interest} + \text{selling price} - \text{buying price}$$
$$= 4800 + 9200 - 10\ 000$$
$$= 4000$$

His \$10 000 investment will have earned \$4000 in 8 years.

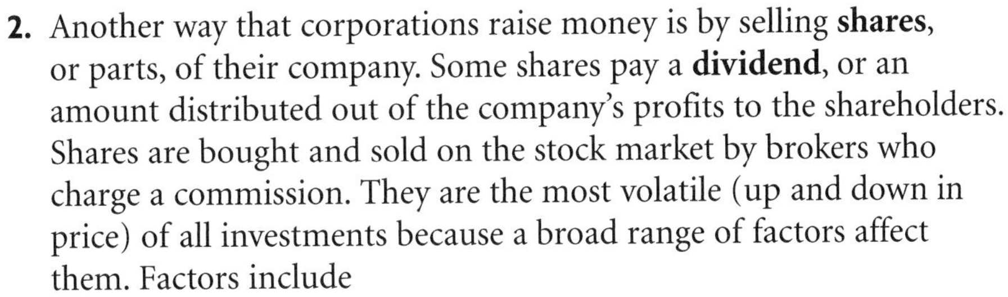

**2.** Another way that corporations raise money is by selling **shares**, or parts, of their company. Some shares pay a **dividend**, or an amount distributed out of the company's profits to the shareholders. Shares are bought and sold on the stock market by brokers who charge a commission. They are the most volatile (up and down in price) of all investments because a broad range of factors affect them. Factors include

- the profitability of the company
- the prospect of future earnings
- the state of the economy
- political crises
- consumer and investor confidence
- interest rates

As a result, the shares you buy will likely be higher or lower in price when you want to sell them.

**a)** How do people earn money with shares?
**b)** When do people lose money with shares?

### Example 2

Mary buys 1000 shares of Acme Corporation at $8.50 per share. She pays $120 commission to buy them. The shares do not pay a dividend.

**a)** If Mary sells the shares for $10.75 per share and pays $134 commission, how much would she have earned buying and selling these shares?

**b)** If Mary had to sell the shares for $7.50 and pay $110 commission, how much would she have lost buying and selling these shares?

### Solution

**a)** increase per share = selling price − buying price
= 10.75 − 8.50
= 2.25

income from sale = increase per share × number of shares
= 2.25 × 1000
= 2250

earnings = income from sale − buying commission − selling commission
= 2250 − 120 − 134
= 1996

Mary would earn $1996.

**b)** decrease per share = buying price − selling price
= 8.50 − 7.50
= 1.00

loss from sale = decrease per share × number of shares
= 1.00 × 1000
= 1000

loss = loss from sale + buying commission + selling commission
= 1000 + 120 + 110
= 1230

Mary would lose $1230.

**3.** Instead of buying bonds or shares of individual corporations, you can invest in a large pool of bonds or shares of different corporations called a **mutual fund.** Units in a mutual fund are bought and sold by the mutual fund company through brokers. As you saw in Chapter 6, banks act as brokers. The value of the unit depends on the total value of all the bonds or shares in the mutual fund. Management fees are charged. Sometimes fees are charged when you buy and sometimes when you sell.

**a)** How do people earn money with mutual funds?

**b)** When do people lose money with mutual funds?

**4.** The advantages of mutual funds are
- diversification (variety of investments)
- professional management
- flexible amounts
- record keeping

The disadvantages of mutual funds are
- management fees and expenses
- loss of control over investment decisions
- possible managers' mistakes

For what types of investors do you think the advantages would outweigh the disadvantages?

## Example 3

Kyle invested \$2000 in a precious metal mutual fund. He paid \$5.73 for each unit. The fee was included in the price.

**a)** How many units, rounded to three decimal places, did he buy?

**b)** Kyle sold the mutual fund units for \$5.87 each. How much did Kyle earn or lose buying and selling these mutual funds?

### Solution

**a)** number of units = amount invested ÷ unit price

$= 2000 \div 5.73$

$\doteq 349.040$

Kyle bought 349.040 units.

**b)** Kyle earned rather than lost because the unit price increased.

increase per unit = selling price − buying price

$= 5.87 - 5.73$

$= 0.14$

earnings from sale = increase per unit × number of units

$= 0.14 \times 349.040$

$\doteq 48.87$

Kyle earned \$48.87.

**5.** Calculate Kyle's earnings from Example 3 using a different method. What do you notice?

## Practise

6. Denver buys a $10 000 ten-year bond from Central Power Corporation. The bond pays 7% simple interest per annum.
   **a)** How much interest will Denver receive each year?
   **b)** If he sells the bond for $9500 after owning it for 7 years, how much will his $10 000 have earned in the 7 years?

7. Sylvia bought 500 shares in Sweet Sue's Chocolate Corporation for $15.10 per share. The commission was $105. The shares do not pay a dividend.
   **a)** How much did Sylvia pay for the shares?
   **b)** Sylvia sold the 500 shares at $15.15 per share. The commission was $97. How much did Sylvia earn or lose buying and selling these shares?

8. Edwina has 1000 shares of Buildem Toy Corporation. The company made a large profit this year and paid a dividend of $1.20 per share. How much does Edwina receive?

9. Kim invests $5000 in Balanced Growth Mutual Fund. The price per unit is $12.05. How many units, rounded to three decimal places, does she buy?

10. Troy has 1524.331 units of Asian Growth Mutual Fund. Units are selling for $5.93 each.
   **a)** What is the value of Troy's units?
   **b)** The fund charges Troy a fee of 0.8% of the value of his account. How much does the fund charge Troy?
   **c)** After the fee is paid out of Troy's account, how many units does Troy have left?

11. Identify the features of each type of investment.
   **a)** bonds **b)** stocks **c)** mutual funds

### Skills Check — Operations with Decimals

Calculate each total value, rounded to the nearest cent.

**a)** 2000 shares at $5.60 per share
**b)** 5000 shares at $0.86 per share
**c)** 12 000 shares at $1.45 per share
**d)** 8000 shares at $2.94 per share
**e)** 693.255 units at $6.80 per unit
**f)** 245.739 units at $8.22 per unit
**g)** 483.440 units at $5.72 per unit
**h)** 523.386 units at $7.29 per unit

# 7.2 Registered Retirement Savings Plans

GICs, mutual funds, bonds, and shares can be held in a **Registered Retirement Savings Plan (RRSP)**.

The amount that you can contribute to an RRSP each year depends upon your income from the year before. Each year that you contribute to an RRSP, your income tax is reduced because your income is reduced by the amount of your contribution. You don't pay income tax on the money that you contribute or on the money it earns until you withdraw it from the plan, usually at retirement.

## Explore

At the age of 20, Jenna decided that she would invest $2000 each year for just 10 years in a GIC inside an RRSP.

At the age of 20, Leah decided that she would not think about RRSPs until she was 30. She would invest $2000 a year in a GIC inside an RRSP the year she was 30 and keep doing the same until she was 65.

How much would each woman invest altogether?

Assume constant interest rates of 8% per annum, compounded semi-annually. Which woman do you think would have more money inside her RRSP the year after she turned 65? Why?

## Develop

**1.** Consider Jenna's investment of $2000 each year from age 20 for just 10 years at a constant interest rate of 8% per annum, compounded semi-annually. This spreadsheet can be used to determine the amount of Jenna's investment the year after she turns 65.

| | A | B | C |
|---|---|---|---|
| 1 | Starting Age (years): | | 20 |
| 2 | Investment per Year for 10 Years ($): | | 2000 |
| 3 | Rate per Annum (%): | | 8 |
| 4 | Compounding Periods in a Year: | | 2 |
| 5 | | | |
| 6 | Age (years) | Principal, P ($) | Amount, A ($) |
| 7 | 20 | 2000.00 | 2163.20 |
| 8 | 21 | 4163.20 | 4502.92 |
| 9 | 22 | 6502.92 | 7033.56 |
| 10 | 23 | 9033.56 | 9770.70 |
| 11 | 24 | 11770.70 | 12731.19 |
| 12 | 25 | 14731.19 | 15933.26 |
| 13 | 26 | 17933.26 | 19396.61 |
| 14 | 27 | 21396.61 | 23142.57 |
| 15 | 28 | 25142.57 | 27194.20 |
| 16 | 29 | 29194.20 | 31576.45 |
| 17 | 30 | 31576.45 | 34153.09 |
| 18 | 31 | 34153.09 | 36939.98 |
| 19 | 32 | 36939.98 | 39954.28 |

**a)** The specifics for Jenna's investment are entered in cells C1, C2, C3, and C4. Why is 2 the value in cell C4?

**b)** How was the value in cell C7 calculated?

**c)** How was the value in cell B8 calculated?

**d)** Because Jenna invests $2000 each year for only 10 years, what do you notice about the values in cells B17, B18, and B19 compared to those before B17?

AT **2.** Examine the full results of the spreadsheet for Jenna's investment.

**a)** How much will Jenna have invested altogether?

**b)** How old will she be when the amount in her RRSP is double the amount in part a)?

**c)** How much will she have in her RRSP the year after she turns 65?

AT **3.** Consider Leah's investment of $2000 each year from age 30 to age 65 at a constant interest rate of 8% per annum, compounded semi-annually. The spreadsheet below can be used to determine the amount of Leah's investment the year after she turns 65.

Unlike the principal for Jenna, who invests $2000 each year for only 10 years, the principal for Leah is the amount from the previous year plus her $2000 investment each year.

| | A | B | C |
|---|---|---|---|
| 1 | Starting Age (years): | | 30 |
| 2 | Investment per Year ($): | | 2000 |
| 3 | Rate per Annum (%): | | 8 |
| 4 | Compounding Periods in a Year: | | 2 |
| 5 | | | |
| 6 | Age (years) | Principal, P ($) | Amount, A ($) |
| 7 | 30 | 2000.00 | 2163.20 |
| 8 | 31 | 4163.20 | 4502.92 |

Examine the full results of the spreadsheet for Leah's investment.

**a)** How much will Leah have invested altogether?

**b)** How old will she be when the amount in her RRSP is double the amount in part a)?

**c)** How much will she have in her RRSP the year after she turns 65?

**4.** Compare your answers to questions 2 c) and 3 c).

**a)** Which woman will have more money the year after she turns 65?

**b)** How much more will she have?

**c)** Do the results surprise you? Is it what you thought when you answered the Explore question on page 133?

AT **d)** If Jenna decided to contribute $2000 each year until she was 65, she would have $951 794.20 the year after she turns 65. Confirm this by using the spreadsheet for Leah. Change the starting age to 20 and FILL DOWN all the columns until the age is 65.

## Practise

AT **5.** Using the spreadsheet for Leah, determine the amounts of the following investments the year after each investor turns 65. Assume constant interest rates of 7% per annum, compounded semi-annually.

**a)** starting at age 20, $1000 each year to age 65

**b)** starting at age 40, $2000 each year to age 65

**c)** starting at age 50, $3000 each year to age 65

**6.** Justify this statement:
Planning for retirement is something to start thinking about early.

# 7.3 Watching Investments

Remember that to **invest** is to set money aside and put it to work in order to get more money back in the future. Investments are long term.

## Explore

This graph shows the performance of Silver Petroleum Corporation shares over a five-year period. If you bought and sold shares of Silver Petroleum in that time period, when would you have preferred to have bought and when would you have preferred to have sold? Explain.

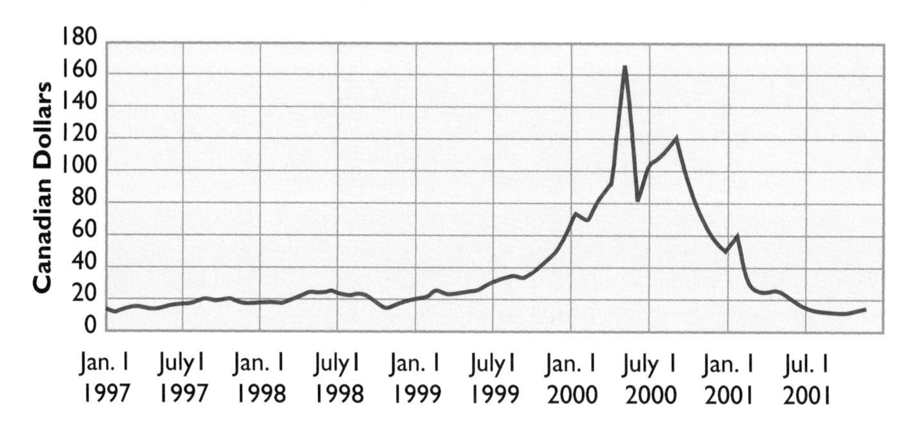


AT

## Develop

Here are a few of the many Canadian companies listed on the Toronto Stock Exchange (TSE). Their symbols follow their names.

| Company | Symbol | Company | Symbol |
|---|---|---|---|
| Abitibi-Consolidated Inc. | A | General Motors Corp. | GM |
| Alcan Inc. | AL | Hudson's Bay Company | HSBC |
| Algoma Central Corp. | ALC | Inco Limited | N |
| Bank of Montreal | BMO | Noranda Inc. | NRD |
| BCE Inc. | BCE | Nortel Networks Corp. | NT |
| Bombardier Inc. | BBD.B | Petro-Canada | PCA |
| Canadian Tire Corp. | CTR | Shell Canada Limited | SHC |
| DaimlerChrysler | DCX | Stelco Inc. | STE.A |

Access to the TSE Web site can be gained through the *Mathematics for Everyday Life 11* page of irwinpublishing.com/students.

**1. a)** Select a company from the list above.

**b)** Look it up on the TSE Web site by typing its symbol in the "QUICK QUOTE" box and clicking "GO."

**c)** Display a graph of its performance over the last five years by locating "Charts" and clicking "5 yr.(m)."

2. Use the graph showing the value of the shares from question 1.
   a) In which six-month period was the price the highest?
   b) Estimate the highest price.
   c) In which six-month period was the price the lowest?
   d) Estimate the lowest price.
   e) Is there any trend evident? If so, describe it.

AT **Practise**

3. a) Select three or four companies from the list above question 1 to track until the end of the term.
   b) Look up each company on the TSE Web site by typing its symbol in the "QUICK QUOTE" box and clicking "GO."
   c) Display a graph of its performance over the last month by locating "Charts" and clicking "1 mo."
   d) Print and post the graph.
   e) Once a week,
      - look up each company on the TSE Web site to check its performance
      - display a graph of each company's performance for the past month, noting particularly the last week
      - print and post the graph beside the previous one
      - create a question about each company for your classmates to solve

**Skills Check** **Mental Math**

Use mental math to find the total value of each.

a) 100 shares at $2.65 per share
b) 1000 shares at $7.14 per share
c) 1000 shares at $13.18 per share
d) 10 000 shares at $9.66 per share
e) 200 shares at $4.20 per share
f) 500 shares at $12.20 per share
g) 2000 shares at $17.13 per share
h) 40 000 shares at $32.10 per share

# 7.4 Risk Tolerance

Investment patterns tend to follow the stages of life. As people's circumstances change, so should how they invest.

We will consider these four stages of life with respect to investing:
- early career
- established
- pre-retirement
- retirement

## Explore

In which of the above life stages do you think people can withstand the most fluctuation (increases and decreases) in their financial resources? Explain.

## Develop

**Risk tolerance** is how much fluctuation in your financial resources you can withstand. It is affected by two key factors:
- the actual amount of money you have after expenses
- your ability to recover if you do suffer losses

For most people, the money they have after expenses increases as they move through their working years. As they near retirement, their ability to recover from losses decreases.

There are three basic **investment objectives**:
- safety of principal (not ending up with less than you started with)
- income (getting regular payments—interest or dividends)
- growth (ending up with considerably more than you started with)

The balance of a person's investment objectives changes with life stages.

An investment can earn a return for you in one of two ways. **Income return** is a regular dividend (from a stock) or interest (from a bond). A **capital gain** occurs when you sell an investment for more than you paid for it. The total return on an investment is the sum of the two returns.

The riskier an investment, the greater the potential return (and the greater the potential loss). This graph show some subtypes of the investments you have seen. There are other subtypes as well.

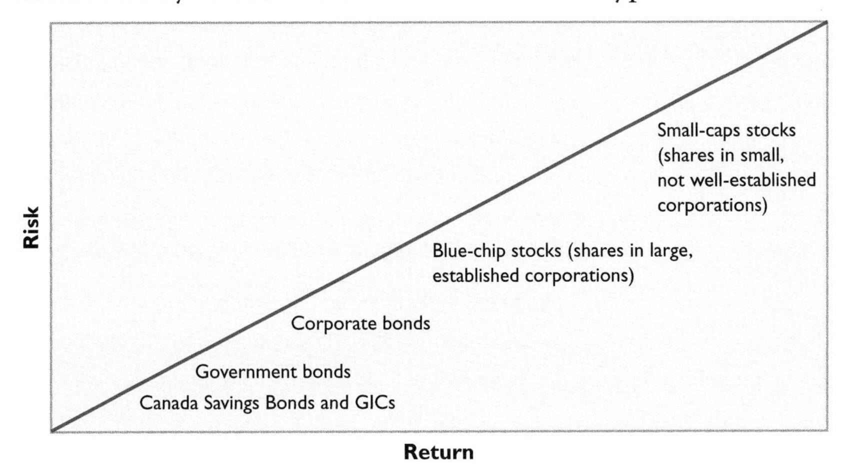


## Example

Alegandro has just retired. He will receive Canada Pension. He must decide what to do with his RRSP investments to supplement his Canada Pension. He knows that a dollar will not likely buy as much in 10 years time as it does today.

**a)** Rank Alegandro's investment objectives. Explain your ranking.

**b)** How would you recommend that he invest the money from his RRSP? Why?

## Solution

**a)** Alegandro's investment objectives might be as follows:

1. safety of principal 2. income 3. growth

Because he is no longer working, Alegandro does not have the opportunity to recover from losses. So, safety of principal is first.

Because he is retired, regular income would supplement his Canada Pension. Therefore, income is second.

Because a dollar will not buy as much in 10 years time as it does today, Alegandro doesn't want his money to lose value. That makes growth third.

**b)** Alegandro might invest about half of the money from his RRSP in Canada Savings Bonds or GICs which are safe and pay interest which would be income. He might invest most of the rest of his money in corporate bonds which will also pay interest as income and have some potential for growth. He might invest a small portion in blue-chip stocks which pay a dividend as income and have more potential for growth.

1. According to the graph, which of each pair is riskier? Why would that be?

   **a)** small-cap or blue-chip stocks **b)** government or corporate bonds

2. When would a capital loss occur?

## Practise

**For each of the following situations, use the investment types shown in the graph on page 139 for your investment recommendations. You can recommend more than one type of investment.**

3. Michelle is 22 years old, single, and working. She has some extra money each month and a reliable car, but would like to own the car of her dreams by the time she is 30.

   **a)** Rank Michelle's investment objectives. Explain your ranking.
   **b)** How would you recommend that she invest her extra money each month? Why?
   **c)** Suppose Michelle forgets the dream car and decides she wants to buy a house. Would your investment recommendation in part b) help her in this scenario? Explain.

4. Meredith is 38 years old. She has been working for more than 15 years and contributing to an RRSP the entire time. She is confident in being able to work for years. Her expenses are under control and she has some surplus money each month. She has accumulated money in GICs and thinks she should diversify, or add variety.

   **a)** Rank Meredith's investment objectives. Explain your ranking.
   **b)** How would you recommend that she invest her surplus money and the money from her GICs? Why?

5. Jason is 21 years old and recognizes the importance of starting an RRSP early. He hopes to contribute the maximum allowed each year.

   **a)** Rank Jason's investment objectives. Explain your ranking.
   **b)** How would you recommend that he invest the money in his RRSP? Why?

6. Louise is 55 years old and her children are grown. She plans to work until she is 65 and she contributes to an RRSP. She has more surplus money each month than ever before, and she has surplus money from selling her home and buying a smaller one. She wants to live comfortably in her retirement.

   **a)** Rank Louise's investment objectives. Explain your ranking.
   **b)** How would you recommend that she invest her surplus money? Why?

# 7.5 Career Focus: Cook

Stanley is a short-order cook in a small restaurant. He is 25 years old, enjoys the business, and hopes to buy it when the owner is ready to retire.

He uses math in his work, particularly when buying food. For example, he needs to consider several patterns when ordering food in quantity for the following breakfast special.

The special is the most popular menu item and the restaurant sells about 500 of them each week.

Each breakfast special includes two slices of toast and either four slices of bacon or three sausages.

About 75% of the breakfast-special orders are with bacon.

Those with white toast and whole wheat toast are about equal.

**1.** Complete this table to determine how much of the five items listed are needed each week for the breakfast special.

| Breakfast Special Orders with | | Number per Order | Total Number Used | Package Size | Number of Packages to Buy |
|---|---|---|---|---|---|
| eggs | | | | 30 eggs per tray | |
| bacon | | | | 20 slices per package | |
| sausages | | | | 144 sausages per box | |
| whole wheat toast | | | | 20 slices per loaf | |
| white toast | | | | 20 slices per loaf | |

**2. a)** Why would they need to buy more of the items than you calculated in question 1?

**b)** What other foods must be bought for the breakfast special?

3. The owner plans to retire in 15 years. Stanley would like to have a sizable down payment to make by then. This past year he saved $2000.
   **a)** If he saved the same amount each year, how much would he have saved?
   **b)** Why should Stanley not just put his money in a savings account?
4. Recall these investment objectives: safety of principal, income, and growth. Rank them for Stanley and explain your ranking.
5. List the advantages and disadvantages of each of the following types of investment for Stanley.
   - GICs
   - Canada Savings Bonds
   - government bonds
   - corporate bonds
   - blue-chip stocks (shares in large, established corporations)
   - small-cap stocks (shares in small, not well-established corporations)
6. Based on your answers to questions 4 and 5, how would you recommend that Stanley invest his money?
7. Stanley's boss plans to use the money from the sale of his business, along with money in an RRSP and his Canada Pension, for his retirement living.
   **a)** Rank his investment objectives when he retires. Explain your ranking.
   **b)** How would you recommend that he invest his money upon retirement? Why?

# 7.6 Putting It All Together: Investing Money

In the previous chapters, you
- found a job
- determined your gross and net pay
- identified your living expenses
- planned to save at least $75 a month
- purchased items
- determined your banking needs
- calculated the amount of a $2000 GIC that you were given and that has now matured

Now you will make plans to invest this $2000 plus interest.

**1.** Refer to your work from Section 6.10. In question 6, how much money did you determine that you had to invest?

**2. a)** What would you like your investment of $2000 plus interest to be earning money towards?

**b)** How much would you like it to amount to before you use it?

**c)** Rank your investment objectives. Explain your ranking.

**3. a)** List the investment options that you have studied.

**b)** For each option in part a), identify which of the following apply.
- a management fee and/or fees for buying and selling to be paid by you
- a commission for buying and selling to be paid by you
- interest to be paid to you
- dividends possibly to be paid to you
- the possibility of a capital gain or capital loss for you

**4. a)** Which investment options from question 3 suit your situation from question 2? Why?

**b)** For how long would you invest the $2000 plus interest? Explain.

# 7.7 Chapter Review

**1.** Bernard bought a $5000 fifteen-year bond issued by Markus Corporation. The bond pays 6% simple interest per annum.

**a)** How much interest will Bernard receive each year?

**b)** If Bernard holds the bond for 15 years, how much interest will he earn?

**c)** If he sells the bond for $5100 after owning it for 5 years, how much will his $5000 have earned for him in 5 years?

**2.** Alvin bought 500 shares of The High Tech Corporation at $5.70 per share. The commission was $45. The shares do not pay a dividend.

**a)** How much did Alvin pay for the shares?

**b)** Alvin sold the 500 shares at $10.50 per share. The commission was $77. How much did Alvin earn or lose buying and selling these shares?

**3.** Tanya invests $1500 in a mutual fund. The price per unit is $9.85. How many units, rounded to three decimal places, did she buy?

**4.** List two advantages and two disadvantages of mutual funds.

**5.** Apply what you have learned about investing in an RRSP, to answer the following.

Which person would have more money by age 65 and why? Assume that both investments earn 6% per annum, compounded annually.

- Trevor invests $1000 in an RRSP starting at age 25 and does so each year until he is 65.
- Reid invests $3000 in an RRSP starting at age 45 and does so each year until he is 65.

**6.** This graph shows the performance of The Marvellous Toy Company shares over a five-year period.

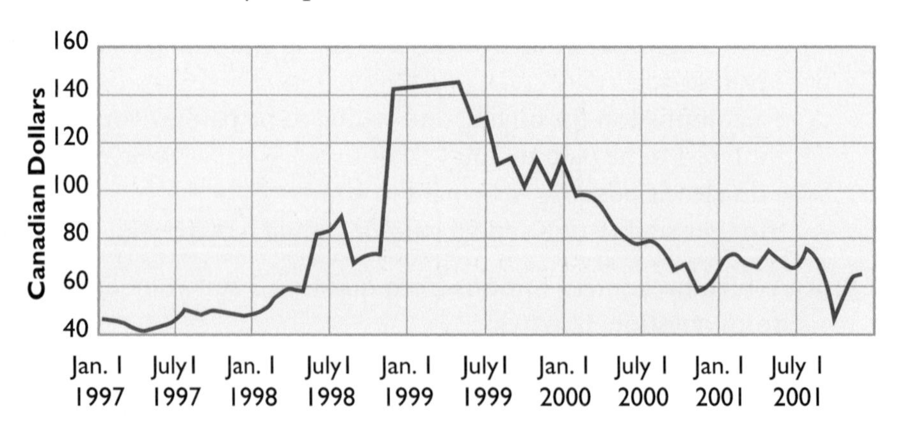

a) In which six-month period was the price the highest?
b) Estimate the highest price.
c) In which six-month period was the price the lowest?
d) Estimate the lowest price.

7. Name two sources of information about stocks and bonds traded on the Toronto Stock Exchange that can be used to track their performance.

8. Identify and explain the two ways to earn a return from an investment.

9. a) Explain what is meant by risk tolerance.
b) What two key factors affect a person's risk tolerance?

10. Tara is 70 years old. She receives Canada Pension every month. She has just sold her house and has $20 000 she wants to invest to supplement her pension income.
a) Rank Tara's investment objectives. Explain your ranking.
b) How would you recommend that she invest the money from the sale of her house? Why?

11. Frank is 39 years old, has a steady job, and has no trouble meeting his expenses. He has some surplus money each month that he would like to invest.
a) What is his life stage?
b) Rank Frank's investment objectives. Explain your ranking.
c) How would you recommend that he invest his surplus money each month? Why?

12. For each of the following options for investing $3000,
- determine what the $3000 will be worth in five years ***or*** explain why it is not possible to determine what it will be worth
- explain why you would or would not recommend it to Kristine, who is 25 years old, with $3000 to invest outside her RRSP, and no immediate plans for her surplus money

a) a GIC earning 4% interest, compounded semi-annually for 5 years
b) a government bond earning 5.5% simple interest each year
c) a corporate bond earning 6% simple interest each year
d) blue-chip stock that can be bought for $12.80 per share and $102 commission, paying $3 per share dividends quarterly
e) small-cap stock that can be bought for $4.90 per share and $105 commission, paying no dividend

CHAPTER

# 8 Taking a Trip

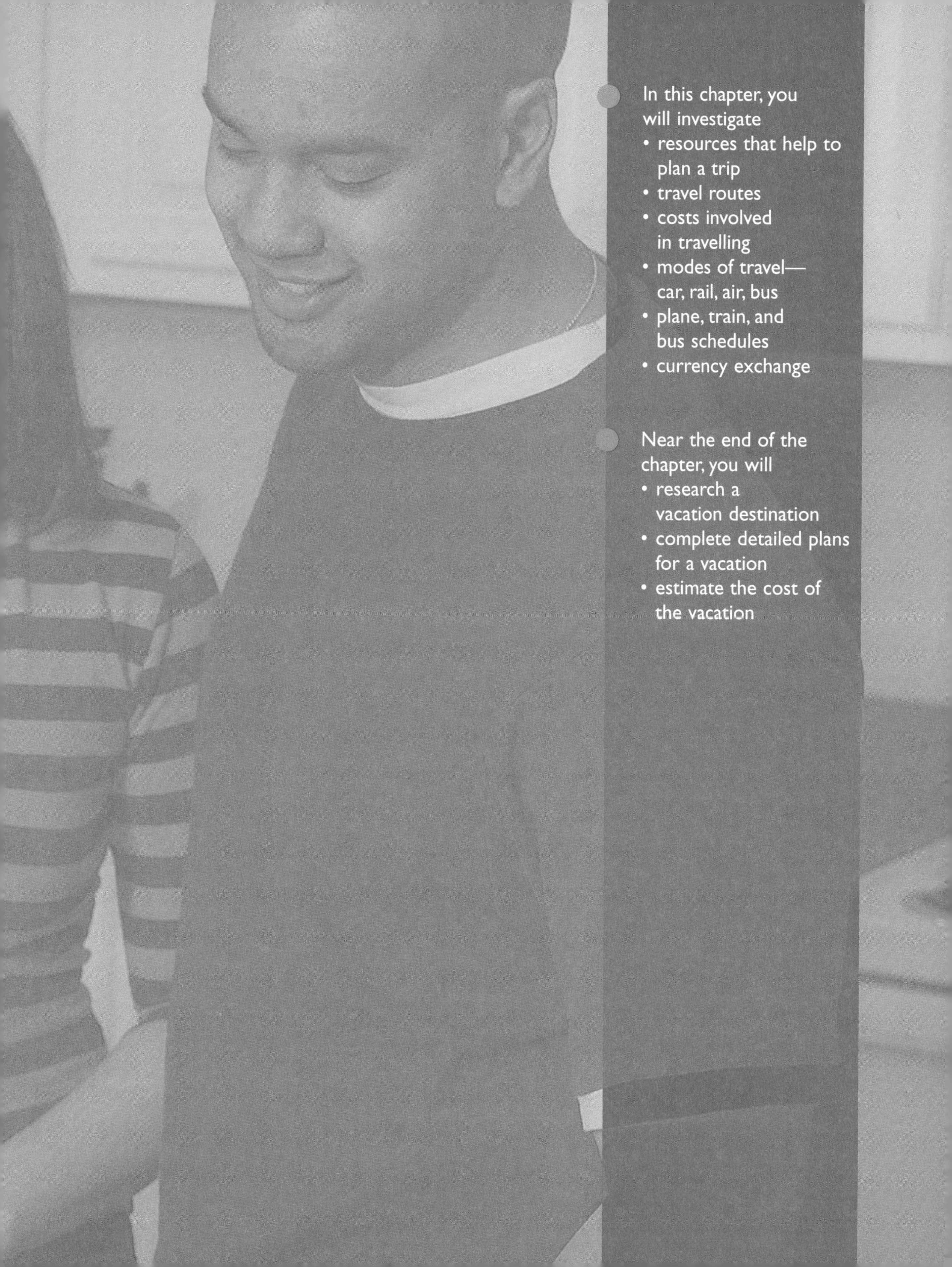

In this chapter, you will investigate

- resources that help to plan a trip
- travel routes
- costs involved in travelling
- modes of travel—car, rail, air, bus
- plane, train, and bus schedules
- currency exchange

Near the end of the chapter, you will

- research a vacation destination
- complete detailed plans for a vacation
- estimate the cost of the vacation

# 8.1 Planning a Car Trip

WS

## Explore

You are planning a driving vacation to the Calgary Stampede next summer. Use road maps to plan a route there and back, naming the highways travelled and the communities along the route.

WS

## Develop

**Questions 1 to 6 are about the route you planned in the Explore question.**

**1.** Use road maps to estimate the driving distance from your home to Calgary and back.

**2. a)** Use road maps to determine
- about how many hours you will drive each day
- about how many kilometres you will drive each day
- where you will stay each night on your way there and back

**b)** If you were driving to Alberta in the winter to ski, what driving factors would be different than those for travelling in the summer?

### Example

Suppose gas costs 72.5¢ per litre and a car consumes 9.7 L of gas per 100 km. How much would the gas cost for a 500 km trip?

### Solution

$$\begin{aligned} \text{fuel consumed} &= \text{distance travelled} \times \text{gas consumption rate} \\ &= 500 \times \frac{9.7}{100} \\ &= 48.5 \end{aligned}$$

The car would consume 48.5 L of gas on a 500 km trip.

$$\begin{aligned} \text{total cost} &= \text{gas consumed} \times \text{cost per litre} \\ &= 48.5 \times 72.5 \\ &= 3516.25 \end{aligned}$$

The total cost was calculated using cents.
3516.25¢ is $35.16, rounded to the nearest cent.

The gas would cost $35.16.

**For the following questions, assume that you will**

- **travel with another person and share expenses**
- **travel to Calgary directly without stops for sightseeing or visiting**
- **stay one week in Calgary**
- **take one side trip out of Calgary to Drumheller to see the dinosaurs at the Royal Tyrrell Museum**

**3.** Use the current cost of gasoline and work with a highway fuel consumption rate of 9.7 L/100 km.

  **a)** Estimate the amount you would expect to spend on gasoline travelling to Calgary and back.

  **b)** Estimate the amount you would expect to spend on gasoline while in Calgary, including driving within the city and taking the side trip to Drumheller.

**For questions 4 to 9, use the Internet, travel books, or brochures.**

**4.** In question 2, you identified where you will be stopping each night on your drive to Calgary and back. Answer the following for each place.

  **a)** Select a type of **accommodation**, such as campground, motel, hotel, or bed and breakfast, for each night.

  **b)** Name two factors that might affect your choice of accommodation.

  **c)** Estimate the cost of accommodations for each place, including applicable taxes.

**5.** Select a type of accommodation at your **destination**, in this case, Calgary. Estimate the cost of accommodations while in Calgary during the Stampede, including applicable taxes. Is it more or less expensive to stay in Calgary at other times of the year?

**6.** Based on your experience of eating out, estimate the cost of the meals while you are travelling and while you are in Calgary and Drumheller.

7. Estimate the cost of admission to the Calgary Stampede for the number of days you plan to attend, including any taxes.

8. Estimate the cost of admission to the Royal Tyrrell Museum, including any taxes.

Access to Web sites about the Calgary Stampede and Royal Tyrrell Museum can be gained through the *Mathematics for Everyday Life 11* page of irwinpublishing.com/students.

9. Estimate the cost of souvenirs and other miscellaneous expenses during the entire trip.

WS **10.** Total the estimates of your expenses. Present your results to the class. Compare your results with those of others. Why do you think there are differences?

11. List the sources of information that you have used to plan the driving vacation to Calgary.

## Skills Check **Mental Math**

If the time is 01:45 in Parry Sound, Ontario, what time is it in

**a)** Calgary, Alberta?
**b)** Saint John, New Brunswick?
**c)** St. John's, Newfoundland?
**d)** Yellowknife, Northwest Territories?
**e)** Moosonee, Ontario?
**f)** Prince Rupert, British Columbia?

You will refer to this map in later sections.

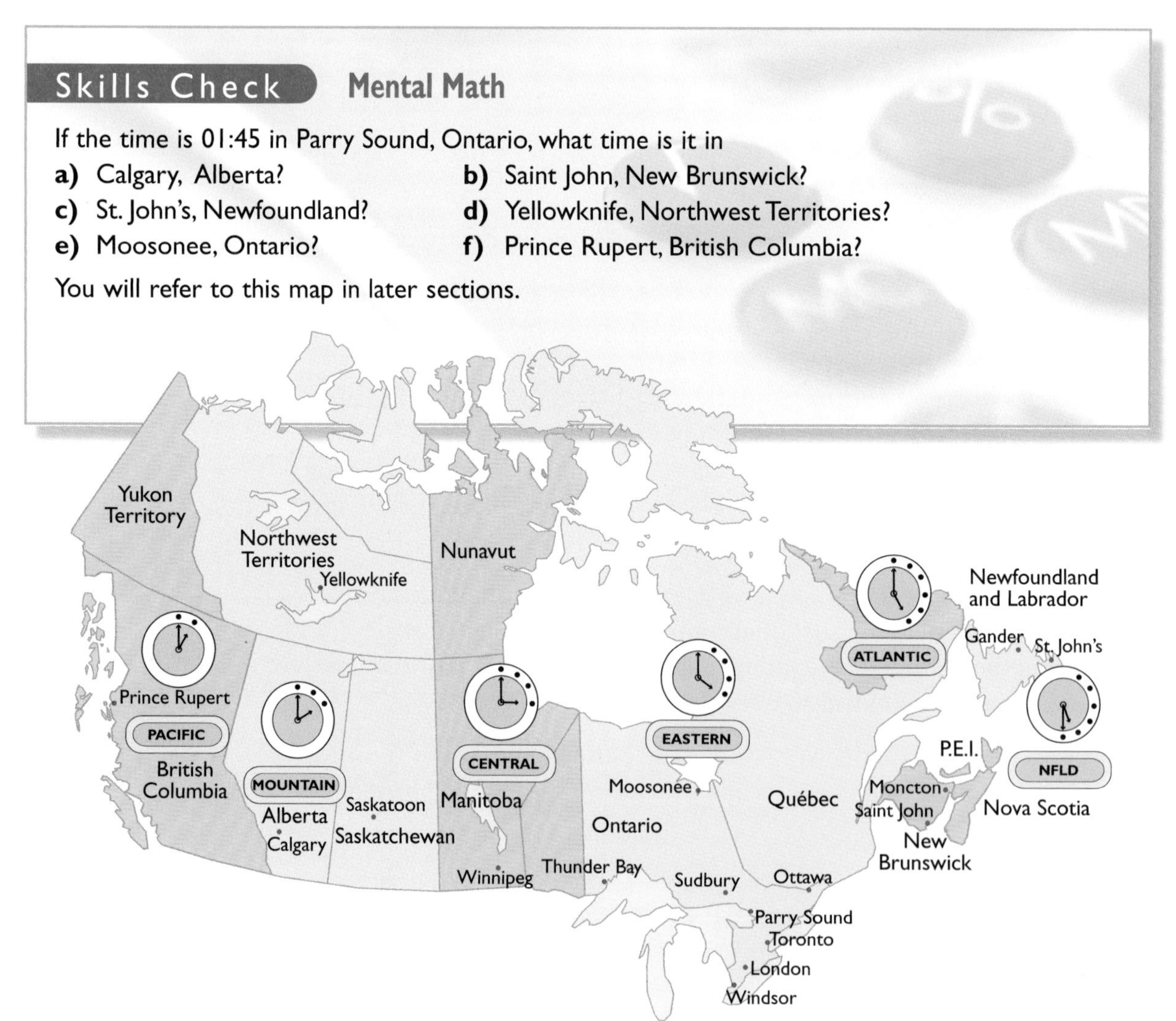

# 8.2 Other Modes of Travel

## Explore

In Section 8.1, you planned a driving trip to Calgary. What other **modes of travel**, or ways of travelling, to Calgary are possible? What are sources of information about routes, schedules, and costs for these modes of travel?

## Develop

**1.** An airline advertises special rates for flights.

**Flights from Toronto**

| Destination | One Way* | Round Trip* |
|---|---|---|
| Ottawa, ON | $188 | $288 |
| Vancouver, BC | $388 | $398 |
| Winnipeg, MB | $168 | $268 |
| Thunder Bay, ON | $140 | $190 |
| Moncton, NB | $288 | $368 |

*Prices do not include $12 security fee each way and 7% GST on total.

Determine the total cost of each flight below.

**a)** a round trip to Winnipeg from Toronto

**b)** a one-way trip to Moncton from Toronto

**2.** Janine compares the costs for travelling by train to London from Belleville.

Train Fares (including taxes)

| | One Way | Return | |
|---|---|---|---|
| Belleville to London | $77.04 | $154.08 | one adult |
| | $60.99 | $107.00 | one adult booked five days in advance |

**a)** What is the most economical way for Janine to book a trip to London?

**b)** Discounts of at least 30% for return trips booked five days in advance are advertised. Does the discount appear to apply to this trip?

3. Tiernan and Gregory want to travel to Pembroke from Peterborough by bus.

Bus Fares (including taxes)

| | One Way | Return | |
|---|---|---|---|
| Peterborough to Pembroke | $51.57 | $103.15 | one adult |
| | $51.57 | $103.15 | two people travelling together, booked at least seven days in advance |

a) What is the most economical way for Tiernan and Gregory to book their trip?

b) The bus company offers a 10% discount for tickets booked on-line. What would be the cost for a regular return trip booked on-line?

4. Why might a person buy a one-way ticket?

## Practise

5. Filipe wants to travel to Ottawa from Toronto to visit his cousin.
   a) Use the flight costs on page 151. Find the total cost of a round-trip flight.
   b) The return bus fare is $95.23, including taxes, and the return train fare when booked in advance is $132, including taxes. Which of the three modes of travel is the least expensive?
   c) What other costs should be factored into each mode of travel?
   d) What factors other than cost might Filipe consider before deciding which way to travel?
   e) What other modes of travel might he consider?

6. Use the Internet, travel brochures, or ads in the travel sections of newspapers to determine the cost, including taxes, of adult return travel by plane, train, and bus between
   a) Sarnia and Toronto
   b) Ottawa and Vancouver
   c) Toronto and Québec City

Access to Web sites with plane, train, and bus fares can be gained through the *Mathematics for Everyday Life 11* page of irwinpublishing.com/students.

7. **a)** For each pair of cities in question 6, compare the cost of the three ways of travelling. Which mode is most expensive? least expensive?
   **b)** Fares often vary. What are some of the conditions that affect them?

8. Use the Internet, travel brochures, or ads in the travel sections of newspapers.
   **a)** Determine the cost, including taxes, of return travel by plane, train, and bus between where you live (or the nearest place from which you can take a plane, train, or bus) and Calgary.
   **b)** Compare the costs in part a) with the cost of gasoline, accommodations, and meals while travelling to Calgary and back that you found in Section 8.1.

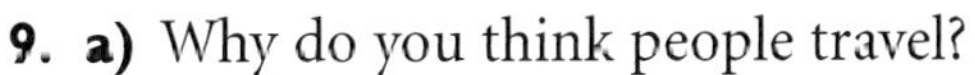

9. **a)** Why do you think people travel?
   **b)** How might the reason for the trip affect the choice of travel mode?

## Skills Check — Mental Math

Times on travel schedules are usually given using the 24-hour clock.

07:28 means 7:28 a.m.
12:56 means 12:56 p.m.
13:15 means 1:15 p.m.
00:15 means 12:15 a.m. (15 minutes after midnight)

Write these times using the 12-hour clock (a.m. or p.m.).

**a)** 18:18 **b)** 9:27 **c)** 12:02 **d)** 00:14 **e)** 23:09

How long is it from

**f)** 2:30 to 12:03? **g)** 12:10 to 15:05? **h)** 20:07 to 23:07?

# 8.3 Reading Schedules

## Explore

This schedule shows summer flight times between Moncton, New Brunswick, and Toronto. Times given on a schedule are **local time**, or the time at the place named.

| Toronto to Moncton | | | |
|---|---|---|---|
| **Day** | **Dates** | **Departure** | **Arrival** |
| Mo–Fr | May 1–Oct 31 | 06:40 | 09:35 |
| Sa–Su | Jun 16–Sep 23 | 14:50 | 17:45 |

| Moncton to Toronto | | | |
|---|---|---|---|
| **Day** | **Dates** | **Departure** | **Arrival** |
| Mo–Fr | May 1–Oct 31 | 20:15 | 21:35 |
| Sa–Su | May 5–Oct 28 | 20:15 | 21:35 |
| Su | Jun 17–Sep 23 | 16:50 | 18:10 |

For a weekday flight, how long does it appear to take to fly to Toronto from Moncton? to Moncton from Toronto? Why is there a difference?

## Develop

### Example

Moira wants to fly to Toronto from Moncton on a Sunday in August. Use the schedule above to answer the following.

**a)** How many flights does Moira have to choose from?
**b)** Which flight leaves in the afternoon?
**c)** What is the *actual* length of the afternoon flight?

## Solution

**a)** "Su" must stand for Sunday. August falls between May 5 and October 28 and between June 17 and September 23. Moira has two flights to choose from.

**b)** 20:15 is the first departure time.

20:15 is 8:15 p.m. (20:15 − 12)

16:50 is the second departure time.

16:50 is 4:50 p.m. (16:50 − 12)

The 16:50 departure is in the afternoon.

**c)** The time from the schedule, that is, from 16:50 to 18:10, is 1 hour 20 minutes.

According to the time zone map on page 150, Moncton is 1 hour ahead of Toronto, so add 1 hour to the time from the schedule.

The flight is 2 hours and 20 minutes long.

**1.** Use the flight schedule on the previous page.

**a)** Liam plans to fly to Moncton from Toronto on a Friday in July and return the following Sunday. What flight times can he take?

**b)** Robert is booked to fly on Wednesday, July 23, from Moncton to Toronto. He is advised to be at the airport 1 hour and 15 minutes before flight time. What time should he be at the airport?

**2.** This schedule shows the trains from Windsor to London.

| Train | Departs Windsor | → Arrives London |
|---|---|---|
| 70 | 02/03/15 (06:00) | 02/03/15 (07:53) |
| 72 | 02/03/15 (09:55) | 02/03/15 (11:36) |
| 76 | 02/03/15 (14:05) | 02/03/15 (15:43) |
| 78 | 02/03/15 (17:30) | 02/03/15 (19:13) |

**a)** What date is this schedule for?

**b)** If Aneela needs to be in London for a meeting at 1:00 p.m., which train should she take from Windsor?

**c)** How long will Aneela be on the train?

**d)** At about what time do you think Aneela should arrive at the Windsor train station to catch the train to her meeting?

**3.** This schedule shows times for buses to Winnipeg, Manitoba, from Thunder Bay, Ontario.

| From: **Thunder Bay, ON** | To: **Winnipeg, MB** |
|---|---|
| **Departure Time** | **Arrival Time** |
| 09:45 | 18:10 |
| 14:05 | 22:40 |
| 23:30 | 07:20 |

**a)** If you took the afternoon bus, would you arrive in Winnipeg in time for the 11:00 p.m. news?

**b)** Find the *actual* length of each bus trip.

**c)** Why do you think there are differences in your answers to part b)?

**4.** This schedule shows the bus service to Toronto from Ottawa.

**OTTAWA - TORONTO: Inter-City Express**

**See Table 746 A-D for Semi-Express & Local Service (Toronto - Peterborough - Ottawa)** — **Read Down**

| Schedule Number | 6201 | 6203 | 6205 | 6131 | 6209 | 6211 | 6115 | 6213 | 6117 |
|---|---|---|---|---|---|---|---|---|---|
| 092703 jas — Effective: 28-Oct-03<br>**746-F**<br>Frequency | Ottawa<br>Toronto | Ottawa<br>Toronto | Ottawa<br>Toronto | Ottawa<br>Peterbor'o<br>Toronto | Ottawa<br>Toronto | Ottawa<br>Toronto | Ottawa<br>Peterbor'o<br>Toronto | Ottawa<br>Belleville<br>Toronto | Ottawa<br>Peterbor'o<br>Toronto<br>Thru Bus |
| E OTTAWA, ON (Terminal) ........... Lv | 07:00 | 09:30 | 11:30 | 13:00 | 14:30 | 16:30 | 18:10 | 18:00 | 00:30 |
| E Kanata (Town Centre)..... Lv | 07:20 | I | I | 13:15 | I | I | 18:30 | I | I |
| | Inter-City Express | Inter-City Express | Inter-City Express | I<br>I<br>I | Inter-City Express | Inter-City Express | Semi Express<br>I | Express<br>via 4168401<br>I | I<br>I<br>I |
| A Toronto, ON, Scarborough (Town Centre) | D | D | D | D | D | D | D | D | D |
| A Toronto...(University Ave. & Queen St.) | D | D | D | D | D | D | D | D | D |
| A TORONTO, ON (Bay St. Term.) ........... Ar | 12:05 | 14:25 | 16:25 | 18:45 | 19:25 | 21:25 | 00:20 | 23:15 | 05:55 |

*All trips on Table 746-F Operate Dly including Hol. Tuesday & Wednesday*

**a)** If you wanted the shortest trip possible, which trip would you choose from Ottawa to Toronto?

**b)** Why do you think some trips take longer than others?

**c)** What departure times are offered before noon on Wednesdays?

## Practise

**Use the Internet or printed schedules to answer questions 5 to 9.**

**5. a)** Find a Wednesday morning flight to Toronto from Ottawa.
**b)** Determine the length of time for that trip.

Access to Web sites with plane, train, and bus schedules can be gained through the *Mathematics for Everyday Life 11* page of irwinpublishing.com/students.

**6. a)** Find a Wednesday morning train to Toronto from Ottawa.
**b)** Determine the length of that trip.

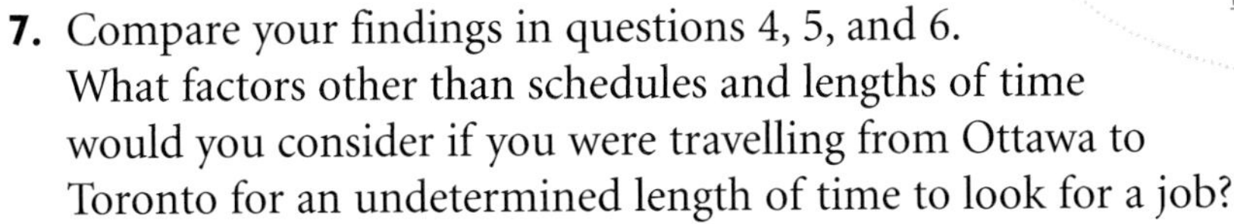

**7.** Compare your findings in questions 4, 5, and 6. What factors other than schedules and lengths of time would you consider if you were travelling from Ottawa to Toronto for an undetermined length of time to look for a job?

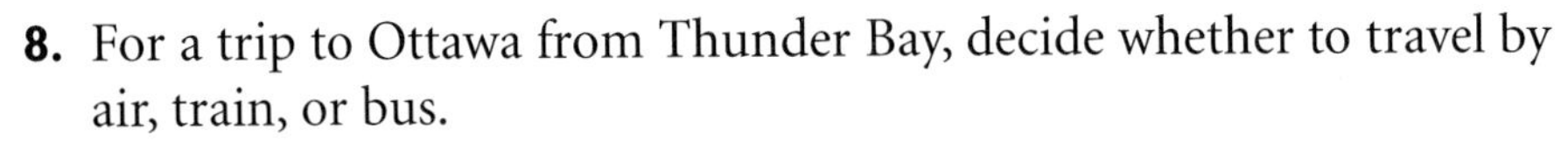

**8.** For a trip to Ottawa from Thunder Bay, decide whether to travel by air, train, or bus.

**a)** Find an afternoon departure time.
**b)** Find the arrival time for that departure.
**c)** Determine the length of that trip.

**9.** Repeat question 8 for a trip to Saskatoon, Saskatchewan, from Sudbury, Ontario.

**10.** Apply what you have learned in Sections 8.1 to 8.3, and discuss the advantages and disadvantages of each mode of travel.

**a)** automobile **b)** plane **c)** train **d)** bus

# 8.4 Travelling Abroad

## Explore

Discuss matters that you should take into account when you travel outside Canada.

## Develop

One of the many things to think about when travelling abroad is using different **currency**, or money.

If you go on a packaged vacation, you might need only spending money in the currency of the country or countries you travel in. If you make your own **itinerary**, or detailed plan of your trip, you will need to pay all expenses as you go.

### Example

Faye sees a gift in London, England, that she'd like to buy for her sister. The price is 12.60 U.K. pounds. Estimate and calculate the price in Canadian dollars. Use the conversion factor (or exchange rate) of 1 U.K. pound = $2.2749 Canadian.

### Solution

**Estimate.**

price in Canadian dollars = price in U.K. pounds × conversion factor
$= 12.60 \times 2.2749$
$\doteq 13 \times 2$
$= 26$

The price of the gift in Canadian dollars is about $26.

**Calculate.** 

price in Canadian dollars = price in U.K. pounds × conversion factor
$= 12.60 \times 2.2749$
$\doteq 28.66$

The price of the gift in Canadian dollars is exactly $28.66.

## Practise

**Use currency exchange rates from the Internet or the newspaper for questions 1 to 5.**

**1.** Lisette and Mariah plan to fly to Denmark to visit relatives.

**a)** What is the name of the currency used in Denmark?
**b)** What is the current value of that currency in Canadian dollars?

**2.** Stephanie is travelling to the Philippines on vacation.

**a)** What is the name of the currency used in the Philippines?
**b)** What is the current value of that currency in Canadian dollars?
**c)** What is the value in Canadian dollars of 4500 units of Philippine currency?

WS **3.** Estimate the cost of each item in Canadian dollars. Then, calculate the cost of each item in Canadian dollars.

**a)** ticket for 2500 Japanese yen
**b)** dinner for 34 euros
**c)** hamburger for 25 Hong Kong dollars
**d)** taxi ride for 70 Israeli new shekels
**e)** sweater for $75 U.S.
**f)** groceries for 250 Polish zlotys
**g)** souvenir for 1200 Russian roubles

Access to Web sites for currency exchange rates and traveller's cheques can be gained through the *Mathematics for Everyday Life 11* page of irwinpublishing.com/students.

**4.** Judy travelled to Australia. She bought a shirt for $17.92 Australian. A similar shirt would cost about $22, including taxes, in Canada. Is it more or less expensive to buy the shirt in Australia?

**5.** James used a credit card to pay 32 U.K. pounds for fish and chips for four people. What charge in Canadian dollars would appear on his credit card statement?

| DATE MONTH/DAY | REFERENCE NO. | PARTICULARS | AMOUNT |
|---|---|---|---|
| 1012 | 01 897255 | ACE GRILL LANCE BAY | 18.76 |
| 1014 | 02 941653 | TC LUGGAGE OTTAWA | 97.30 |
| 1020 | 03 976682 | THE FISH & CHIPS SHOPPE LONDON 32 U.K. POUNDS @ ▬ | ▬ |

**6.** Use the exchange rate 81.17 Japanese yen = $1 Cdn. A particular digital stereo component costs $750 in Canada. What is the most this item could cost in Japan to be a good buy for a Canadian travelling in Japan? What assumptions did you make?

**7.** Traveller's cheques can be used to pay for items in other countries. Research traveller's cheques. What are they? Identify advantages and disadvantages of using them rather than cash, credit card, or debit card.

# 8.5 Career Focus: Flight Attendant

The tourism industry offers a variety of job opportunities. It needs workers who are people-oriented and interested in meeting the needs of others in personal and creative ways.

Janey works as a flight attendant. Her son, Taylor, is interested in the same career.

Flight attendants are often thought of as the people who serve food and beverages on flights. However, they also play key roles in promoting safety. Airlines provide intensive training for flight attendants, including ongoing training in emergency procedures.

1. a) What is one of the major concerns that flight attendants have while doing their jobs?
   b) How do airlines prepare flight attendants to deal with safety and emergency issues?
   c) List four qualities that a flight attendant candidate would probably need to possess to be hired.

2. Recently Janey has been flying between Toronto and Fort Myers, Florida, and back. Use a flight schedule to determine the length of time she is in the air for a round trip.

3. Flight attendants often sell items from in-flight boutiques. Calculate the cost of each item in Canadian dollars. Use currency exchange rates from the Internet or the newspaper.
   a) On a flight from Mexico, Tien bought a toy bear for 85.81 Mexican new pesos.
   b) On a flight from China, Maria bought a watch for 394 Chinese renminbis.
   c) On a flight from the Bahamas, Adrian bought earrings for $18.38 Bahamian.
   d) On a flight from Ireland, Jessica bought a bracelet for 32 euros.

Access to Web sites for flight schedules and currency exchange rates can be gained through the *Mathematics for Everyday Life 11* page of irwinpublishing.com/students.

# 8.6 Putting It All Together: Planning Your Trip

**A** Apply what you have learned in this chapter to plan a one-week vacation to a destination outside of Canada. Determine a reasonable cost, in Canadian dollars, for the entire vacation, including

- transportation
- accommodations
- meals
- other expenses
- all taxes and fees

# 8.7 Chapter Review

**1. a)** Use a road map to plan a route between Timmins and Orillia. Name the highways travelled and the places along the route.
**b)** Determine the distance of the trip along your route.
**c)** At a cost of 70.9¢/L of gas and highway gas consumption of 9.2 L/100 km, how much would the gas cost?
**d)** If you planned to drive that distance in one day, how many hours would you need? Explain.
**e)** How many meals would you require along the way? Estimate the cost.
**f)** How would the time of year affect your travel plans?

**2.** Describe a situation in which you might choose to
**a)** drive rather than fly to a destination
**b)** fly rather than drive to a destination
**c)** take the train to a destination
Explain your reasoning.

**3.** The following are advertised flight costs for departures from Ottawa.

| Destination | Round Trip* |
|---|---|
| Montréal, QC | $198 |
| Calgary, AB | $489 |
| The Pas, MB | $825 |
| Gander, NF | $699 |

*Prices do not include $24 security fee and 7% GST on total.

Calculate the total cost of a round-trip flight from Ottawa to
**a)** Montréal **b)** The Pas **c)** Gander

**4.** Jason is going to Montréal from Ottawa for a long weekend.
**a)** What was the airfare from question 3 a)?
**b)** The return bus fare is $60.56, including taxes, and the return train fare (booked in advance) is $63.13, including taxes. Which of the three modes of travel is the least expensive?
**c)** What other costs should be factored into each mode of travel?
**d)** What factors other than cost might Jason consider before deciding which way to travel?

5. The following shows times for flights between Toronto and Gander, Newfoundland.

| Toronto to Gander | | | |
|---|---|---|---|
| **Day** | **Dates** | **Departure** | **Arrival** |
| Su | Jun 24–Sep 2 | 15:00 | 19:51 |

| Gander to Toronto | | | |
|---|---|---|---|
| **Day** | **Dates** | **Departure** | **Arrival** |
| Su | Jun 24–Sep 2 | 20:40 | 23:55 |

**a)** On what day of the week are the flights offered between Toronto and Gander?
**b)** Explain why the flights might operate on only one day of the week.
**c)** State the local times at which the flight from Toronto departs and arrives, using a.m. or p.m.
**d)** State the local times at which the flight from Gander departs and arrives, using a.m. or p.m.
**e)** What is the *actual* length of the flight from Gander to Toronto?

6. Find the value of each item in Canadian funds.
**a)** admission to a theme park for $49 U.S.
**b)** a scarf for 1000 Japanese yen
**c)** a train trip for two for 75 U.K. pounds

7. Choose a place in Ontario that you would like to visit. Determine possible modes of travel for getting there from where you live. Describe a car route and estimate the costs and time involved for travelling to your destination by car. Discuss the advantages and disadvantages of all possible modes of travel to this destination.

CHAPTER

# 9 Borrowing Money

In this chapter you will

- investigate the features of credit cards
- calculate the interest costs that can be associated with using a credit card
- investigate the features of short-term loans
- calculate the interest costs of short-term loans
- discuss the advantages and disadvantages of borrowing
- investigate the effects of varying how often payments are made

Near the end of the chapter, you will

- choose a method of borrowing to pay for the vacation that you planned and estimated the cost of in Chapter 8
- calculate the interest cost of borrowing to pay for the vacation

# 9.1 Credit Cards

## Explore

What is the difference between paying with debit card and paying with cash? What is the difference between paying with a debit card and paying with a credit card?

## Develop

**1.** Louise has been earning money at an after-school job for more than a year. She is applying for a credit card. She has completed part of the application.

**1. Help us get to know you**

Language preferred: English ✓ French O

First Name/Middle Initial/Last Name: LOUISE A SAYER

Current address: 104 LANE ROAD Apt.#

Number Street/RR/PO.box

City/Town: SMALLTOWN Province: ON Postal Code: K7H1G6

How long at current address?: 10 04 (Years months)

Home Telephone No.: 613 555-8226 Business Telephone No.

E-mail address: lsayer@irwin.com

Previous address (if you have lived at your current address less then 3 years) Apt.#

Number Street/RR/PO.box

City/Town Province Postal Code

How long at previous address? (Years months)

Date of birth: 101783 (M M D D Y Y)

Social Insurance Number: 422-312-651

**2. Tell us about what you do for a living**

Employed O Student ✓ Retired O Homemaker O Other O Unemployed O Self employed** O

**(you may be required to submit 2 years of Revenue Canada Notice of Assessments)

Name of your current employer (if applicable): TRILITE INC

How long with current employer: 01 Years 02 months

**3. Give us a snapshot of your finances**

Your primary annual income $ 4000.00

Housing status: rent O own O live with parents ✓ other O

Monthly mortgage/rent $ .00

Name of Bank/Institution

O Primary Savings Account

✓ Primary Chequing Account: CANSAVE

I, the applicant, accept your invitation to apply for the Credit Card. I have read the terms and conditions on the back of this application and agree to be bound by them.

Date: 050602 (M M D D Y Y) Applicant's signature: *Louise Sayer*

**a)** When is an applicant required to complete the "Previous address" section?

**b)** Why did Louise not complete that section?

**2.** Explain why you think issuers of credit cards want the following information.

**a)** current address
**b)** previous address
**c)** SIN
**d)** employment status
**e)** annual income
**f)** housing status
**g)** bank account numbers
**h)** signature

3. Using a credit card allows you to buy items without paying for them immediately. A credit card is considered a way of borrowing money for a short period of time. What other ways of making purchases and paying for them later do you know? (They were discussed in Chapter 5.)

4. Some stores have their own credit cards that you can use to make purchases at their stores. Credit cards that can be used at a wide variety of stores around the world are issued by banks. There are three main bank-issued credit cards. Do you know what they are? If so, name them.

## Practise

**Use the following information to answer questions 5 to 9.**

Many of the bank-issued cards are available in different types. Examples of three types are outlined below.

**Type A**

Lowest rate of interest available on any Canadian bank credit card, guaranteed!

- Annual fee: none
- Minimum household income required: $15 000
- Minimum student income required: $1200
- Credit limit: $500

**Type B**

A card that rewards you with points for dollars spent or cash back at the end of the year

- Annual fee: $29
- Minimum household income required: $25 000
- Minimum student income required: $2300
- Credit limit: $1500

**Type C**

A prestigious card that rewards you with even more points for dollars spent or cash back at the end of the year and offers other benefits, such as insurance on purchases and travel insurance

- Annual fee: $99
- Minimum household income required: $35 000
- Minimum student income required: $5000
- Credit limit: $5000

5. An **annual fee** is the amount paid once a year that keeps the credit card account open.
   a) What is the annual fee for each type of card?
   b) Why do you think the annual fees vary?

6. a) Why do you think there is a minimum income required?
   b) Why do you think the minimum income required is lower for students?
   c) Larry, a student, earned $9.75 per hour and worked 15 hours a week for 45 weeks last year. For which card(s) is he eligible?

7. **Credit limit** is the maximum amount that you are allowed to have owing on the card at any given time. You can request an increase in your credit limit. If you do, the card issuer checks to see if you owe money on other credit cards and loans before granting an increase. Why would people want higher credit limits?

8. Suppose you plan to pay your credit card bill in full each month when it is due. No interest will be charged. Which card would you choose? Justify your choice.

9. Suppose you plan to occasionally make purchases that you will be unable to pay for completely each month. Interest will be charged. Which card would you choose? Justify your choice.

10. a) Some credit card applications require your mother's maiden name. Why do you think the card issuer wants this information?
    b) Some applications ask if you would like additional cards for the same credit card account. Describe a situation when someone may request an additional card.

WS 11. Use brochures or the Internet to determine the annual fees, income requirements, credit limit, annual rate of interest charged, and other distinguishing features of a bank-issued credit card. Compare your findings with those of classmates who researched different cards.

Access to Web sites about credit cards can be gained through the *Mathematics for Everyday Life 11* page of irwinpublishing.com/students.

## Skills Check — Operations with Decimals

Calculate and round each answer to two decimal places.

a) $1500 \times \left(1 + \frac{0.165}{2}\right)^{(2 \times 3)}$

b) $8000 \times \left(1 + \frac{0.186}{12}\right)^{(12 \times 2)}$

c) $2000 \times \left(1 + \frac{0.12}{4}\right)^{(4 \times 5)}$

d) $250 \times \left(1 + \frac{0.205}{365}\right)^{\left(365 \times \frac{1}{12}\right)}$

# 9.2 Delaying Payments on Credit Card Purchases

When you use a credit card to make a purchase, you are borrowing money for a short period of time. Whenever you borrow money, there is a cost. This cost is interest.

## Explore

Suppose you owe $500. How much interest would you pay in one year at a current credit card interest rate?

## Develop

**1.** Brendan made a purchase on his credit card for $150. He had no **previous balance**, or amount unpaid from the month before. His **new balance**, or amount now owed, is $150. When the payment comes due, he is able to make only the **minimum payment** of $10.

For the next month, his previous balance is how much he still owes.

$$\begin{aligned}\text{previous balance} &= \text{new balance (from month before)} - \text{minimum payment made}\\ &= 150 - 10\\ &= 140\end{aligned}$$

He must pay interest at 18.6% per annum, compounded daily, for one month on the previous balance. Using the compound interest formula, the amount is as follows.

$$\begin{aligned}A &= P(1 + i)^n\\ &= 140 \times \left(1 + \frac{0.186}{365}\right)^{\left(365 \times \frac{1}{12}\right)}\\ &\doteq 142.19\end{aligned}$$

Since the exact month is not known, $\frac{1}{12}$ is used.

Brendan does not make any more purchases, so the amount he owes the next month, the new balance, is $142.19.

Since $A = P + I$, then $I = A - P$.

$$\begin{aligned}I &= A - P\\ &= 142.19 - 140\\ &= 2.19\end{aligned}$$

The interest he pays the first month is $2.19.

**a)** Why was $\frac{0.186}{365}$ substituted for $i$?

**b)** Why was $365 \times \frac{1}{12}$ substituted for $n$?

2. Brendan, from question 1, must make the minimum payment of $10 the next month.

   a) Determine the previous balance on which he pays interest.
   b) Determine his new balance assuming he does not make any more purchases.
   c) Determine the interest he pays.

3. Brendan, from questions 1 and 2, continues to make only the minimum payment of $10 each month and doesn't make any more purchases. This spreadsheet shows all of the interest Brendan pays.

| | A | B | C | D | E |
|---|---|---|---|---|---|
| 1 | Purchase Amount ($): | 150.00 | Total Interest Paid ($): | 19.22 | |
| 2 | Annual Interest Rate (%): | 18.6 | | | |
| 3 | Minimum Payment ($): | 10.00 | | | |
| 4 | | | | | |
| 5 | Month | Previous Balance ($) | New Balance ($) | Interest ($) | Minimum Payment ($) |
| 6 | 1 | | 150.00 | | 10 |
| 7 | 2 | 140.00 | 142.19 | 2.19 | 10 |
| 8 | 3 | 132.19 | 134.25 | 2.06 | 10 |
| 9 | 4 | 124.25 | 126.19 | 1.94 | 10 |
| 10 | 5 | 116.19 | 118.01 | 1.81 | 10 |
| 11 | 6 | 108.01 | 109.70 | 1.69 | 10 |
| 12 | 7 | 99.70 | 101.25 | 1.56 | 10 |
| 13 | 8 | 91.25 | 92.68 | 1.43 | 10 |
| 14 | 9 | 82.68 | 83.97 | 1.29 | 10 |
| 15 | 10 | 73.97 | 75.12 | 1.16 | 10 |
| 16 | 11 | 65.12 | 66.14 | 1.02 | 10 |
| 17 | 12 | 56.14 | 57.02 | 0.88 | 10 |
| 18 | 13 | 47.02 | 47.75 | 0.73 | 10 |
| 19 | 14 | 37.75 | 38.34 | 0.59 | 10 |
| 20 | 15 | 28.34 | 28.78 | 0.44 | 10 |
| 21 | 16 | 18.78 | 19.08 | 0.29 | 10 |
| 22 | 17 | 9.08 | 9.22 | 0.14 | |

   a) Compare your answers to question 2 parts a), b), and c) with the values in cells B8, C8, and D8.
   b) In which month was the new balance less than the minimum payment? How much would his final payment that month be?
   c) How many months did he make the minimum payment?
   d) How much interest does Brendan pay on the $150 purchase?
   e) Check: Do 16 minimum payments of $10 plus one payment of $9.22 equal the $150 purchase plus the $19.22 total interest paid?

## Practise

AT **Use the same spreadsheet as in question 3 to answer questions 4 to 6.**

**4.** Jose purchases a $250 jacket on his credit card. He makes no other purchases. He makes the minimum payment of $12 each month and pays 19.2% per annum compounded daily.

**a)** Enter the values for this purchase in cells B1, B2, and B3. All the values in the columns will adjust.

**b)** Is the last new balance less than the minimum payment? If not, FILL DOWN several rows all the columns, starting with the last row that has a value in each cell.

**c)** Select and delete the values in the rows below the first new balance that is less than the minimum payment.

**d)** Select and delete the value in the "Minimum Payment ($)" column to the right of the now last new balance. Why should you do this?

**e)** How many minimum payments are made?

**f)** How much is the final payment?

**g)** How much interest is paid altogether?

**h)** Check: Do all the minimum payments plus the final payment equal the purchase plus the total interest paid?

**5.** Maria purchases a stereo for $850 on her credit card. She makes no other purchases. She makes the minimum payment of $25 each month and pays 16.1% per annum compounded daily. Repeat parts a) to h) of question 4 for Maria's purchase.

**6.** Rama purchases a new computer system for $2399 on her credit card. She makes no other purchases. She makes the minimum payment of $65 each month and pays 15.5% per annum compounded daily. Repeat parts a) to h) of question 4 for Rama's purchase.

***The calculations that you have made are simplified compared to real credit card interest calculations. Credit card interest is actually charged from the day of purchase. As well, people seldom make only one purchase on a credit card over a period of several months.***

## 9.3 Short-Term Borrowing

Most people need to borrow money for large purchases, such as homes. When money is borrowed to pay for real estate (land with or without buildings), it is called a **mortgage**. The real estate is the **security**, or what the bank can take if payments are not met. The money is usually paid back over a long term. Most other borrowing works the same way, but is short term.

### Explore

How much do you think it costs to borrow $30 000 for 5 years at 6% per annum compounded monthly? Why do you think that?

### Develop

**1.** A **loan**, or money borrowed with a promise to repay, is what many people use for short-term borrowing. People seek a loan to buy something specific, such as a car, business equipment, home improvements, a vacation, or an investment.

Another reason some people seek a loan is to consolidate debt. That means they take out a new loan to pay off existing loans and credit card charges.

The security for a loan can be such things as real estate, business equipment, and future wages. The more valuable the security, the lower the interest rate charged.

**a)** How could you find the current rates for loans?
**b)** Find the current rates.

**2.** A loan is repaid in regular instalments. Payments are made weekly, biweekly, semi-monthly, or monthly until the loan is repaid. However, most loans can be repaid fully or partially at any time. Why might some borrowers choose to make loan payments weekly and others, monthly?

**3.** Banks also offer **lines of credit**. A line of credit is a pre-approved loan without a specific purpose. The borrower has a credit limit, just as with a credit card. Explain what a credit limit is.

**4.** People seek a line of credit to have quick access to money in case they need it in the future. The security for a line of credit can be the same sort of thing that is used for other loans. As with other loans, the more valuable the security, the lower the interest rate charged.

**a)** How could you find the current interest rates for lines of credit?
**b)** Find the current interest rates.

**c)** Are the current interest rates for lines of credit more or less than the current interest rates for credit cards?

**d)** Which way of borrowing money—a line of credit or a credit card—do you think is wiser? Explain.

**5.** You can pay for things using a line of credit by withdrawing money from an ATM, using a debit card, or writing a cheque. You can repay the money you borrow at any time partially or fully, but you must make a minimum monthly payment. Where else have you seen minimum monthly payments required?

**6. a)** How is a line of credit different from other loans?

**b)** What features of a line of credit are similar to features of borrowing on a credit card?

**7.** Banks also offer **overdraft protection**, which means if you withdraw more from your account than you have in it, the bank covers the difference up to an agreed-upon amount. You can repay the money at any time partially or fully, but you must make a minimum monthly payment. The interest charged is comparable to credit card interest rates. Why would people seek overdraft protection?

## Practise

**8.** Apply what you have learned about short-term loans to explain why credit cards and overdraft protection have higher interest rates than other loans.

**9.** Discuss the advantages and disadvantages of each.

**a)** delaying payments on credit card purchases

**b)** using loans to pay for such items as cars, home improvements, and vacations

**c)** borrowing on a line of credit

**d)** having overdraft protection

**10.** Consider current interest rates. Which has higher interest rates—borrowing money or saving money? Why would that be?

### Skills Check — Mental Math

Mentally calculate the number of payments.

**a)** monthly for 5 years
**b)** semi-annually for 7 years
**c)** quarterly for 12 years
**d)** semi-monthly for 3 years
**e)** weekly for 4 years
**f)** biweekly for 6 years

# 9.4 Repaying Loans

## Explore

The **cost of a loan**, that is, the total interest paid, varies depending upon the following factors:

- the amount of the loan
- the length of the loan
- the amount of the regular payments
- the rate of interest
- the number of regular payments made in a year

Which factors above increase the cost of a loan as they become greater?

Which factors decrease the cost of a loan as they become greater?

Explain.

## Develop

**1.** Determine the cost of each loan.

**a)** Ramiro borrows \$18 500 to buy a car and makes monthly payments of \$430 for four years.

**b)** Donald borrows \$50 000 to buy a machine for his business. He makes monthly payments of \$970 for five years.

**2.** An **amortization schedule** is a detailed listing of

- the equal payments required to repay a loan
- the portion of each payment that is interest
- the portion of each payment that is principal
- the balance remaining after each payment is made

Refer to the amortization schedule for Alexis's car loan on the next page.

**a)** The values in cells E3, E4, E5, and E6 are the specifics that are entered for a particular loan. What are the specifics for the car loan?

**b)** The spreadsheet calculates the amount of each payment and displays it in cell E8. It is an involved calculation. What amount did the spreadsheet calculate for each payment for the car loan?

**c)** The spreadsheet calculates the number of payments and displays it in cell E9. How was 36 payments for the car loan calculated?

**d)** As well as generating the list of payment details, the spreadsheet calculates the cost of the loan, or the total interest paid, and displays it in cell E10. How was the loan cost of \$790.24 calculated for the car loan?

3. Refer to the amortization schedule for Alexis's car loan below.
   a) How many years is it going to take Alexis to pay for the car?
   b) How much will the car have cost her?
   c) How often are her payments made?
   d) What do you notice about the portion of each payment that is interest as more payments are made?
   e) What do you notice about the portion of each payment that is principal as more payments are made?

| | A | B | C | D | E |
|---|---|---|---|---|---|
| 1 | Amortization Schedule for a Loan | | | | |
| 2 | | | | | |
| 3 | Number of Payments per Year: | | | | 12 |
| 4 | Annual Interest Rate (%): | | | | 1.9 |
| 5 | Amount of Loan ($): | | | | 26731.73 |
| 6 | Length of the Loan (years): | | | | 3 |
| 7 | | | | | |
| 8 | Spreadsheet Calculation of Payment ($): | | | | 764.50 |
| 9 | Spreadsheet Calculation of Number of Payments: | | | | 36 |
| 10 | Cost of the Loan or Total Interest Paid ($): | | | | 790.24 |
| 11 | | | | | |
| 12 | Payment Number | Payment ($) | Interest ($) | Principal ($) | Balance ($) |
| 13 | | | | | 26731.73 |
| 14 | 1 | 764.50 | 42.33 | 722.17 | 26009.56 |
| 15 | 2 | 764.50 | 41.18 | 723.32 | 25286.24 |
| 16 | 3 | 764.50 | 40.04 | 724.46 | 24561.78 |
| 17 | 4 | 764.50 | 38.89 | 725.61 | 23836.17 |
| 18 | 5 | 764.50 | 37.74 | 726.76 | 23109.41 |
| 19 | 6 | 764.50 | 36.59 | 727.91 | 22381.50 |
| 20 | 7 | 764.50 | 35.44 | 729.06 | 21652.44 |
| 21 | 8 | 764.50 | 34.28 | 730.22 | 20922.22 |
| 22 | 9 | 764.50 | 33.13 | 731.37 | 20190.85 |
| 23 | 10 | 764.50 | 31.97 | 732.53 | 19458.32 |
| 24 | 11 | 764.50 | 30.81 | 733.69 | 18724.63 |
| 25 | 12 | 764.50 | 29.65 | 734.85 | 17989.78 |
| 26 | 13 | 764.50 | 28.48 | 736.02 | 17253.76 |
| 27 | 14 | 764.50 | 27.32 | 737.18 | 16516.58 |

| | | | | | |
|---|---|---|---|---|---|
| 41 | 28 | 764.50 | 10.81 | 753.69 | 6072.65 |
| 42 | 29 | 764.50 | 9.62 | 754.88 | 5317.76 |
| 43 | 30 | 764.50 | 8.42 | 756.08 | 4561.68 |
| 44 | 31 | 764.50 | 7.22 | 757.28 | 3804.41 |
| 45 | 32 | 764.50 | 6.02 | 758.48 | 3045.93 |
| 46 | 33 | 764.50 | 4.82 | 759.68 | 2286.25 |
| 47 | 34 | 764.50 | 3.62 | 760.88 | 1525.37 |
| 48 | 35 | 764.50 | 2.42 | 762.08 | 763.29 |
| 49 | 36 | 764.50 | 1.21 | 763.29 | 0.00 |

## Practise

**Modify the Amortization Schedule for a Loan spreadsheet to answer the following questions.**

**4.** Predict an increase or a decrease in
- the amount of each payment
- the cost of the loan

if each of the following changes is made to Alexis's original loan from question 2.

**a)** The interest rate was 7.8% rather than 1.9%.
**b)** The amount of the loan was $2000 less.
**c)** Alexis made payments weekly rather than monthly.
**d)** The length of the car loan was 4 years rather than 3 years.

**5.** Make the changes to the spreadsheet for each part of question 4. Return to Alexis's original loan each time except for the one different factor. Compare the results with your predictions.

**6.** Kyle borrows money to pay for a new personal watercraft.
- The amount of the loan is $9750.
- The payments are monthly.
- The annual rate of interest is 9.9%.
- The length of the loan is 3 years.

**a)** Enter the specifics for this loan in the appropriate cells.
**b)** Delete the entries, or FILL DOWN the columns so that the list goes to exactly 36 payments. The balance should now be $0.00.
**c)** How much will the personal watercraft cost Kyle by the time he finishes repaying the loan?

**7.** Predict and then use the spreadsheet to show the effect of changing how often payments are made—monthly, semi-monthly, biweekly, or weekly—on the amount of each payment and the cost of the loan for Kyle's loan in question 6. Which change had the greatest impact on reducing the cost of the loan—monthly to semi-monthly, semi-monthly to biweekly, biweekly to weekly?

**8.** Avis takes out a loan to purchase equipment for her business.
- The amount of the loan is $40 000.
- The annual rate of interest is 8.5%.
- The length of the loan is 5 years.

**a)** How often would you suggest that Avis make her payments so that the cost of the loan is as low as possible?
**b)** Determine the amount of each payment and the cost of the loan for the number of payments per year from your suggestion in part a).

# 9.5 Career Focus: Small Business Ownership—Yard Maintenance

Lance and Rita are starting a new business. They plan to do yardwork for homeowners in the spring, summer, and fall.

Customers can choose from the following spring-through-fall yardwork services:

- trimming of shrubs and small trees, and removal of trimmings
- trimming of hedges and removal of trimmings
- removal of fallen leaves
- lawn fertilizing
- lawn aerating
- lawn mowing and mulching
- lawn trimming
- grass seeding
- delivery and laying of sod
- sod removal
- delivery and spreading of top soil
- cultivating of gardens
- minor fence repairs

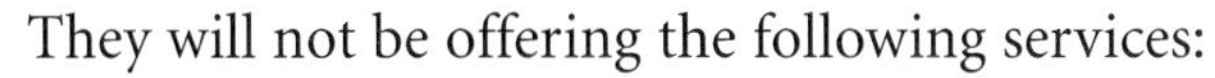

They will not be offering the following services:

- weed control on lawns
- vegetation control on walks and driveways
- grass clippings removal

They have a pickup truck on which there is still a year of monthly payments to be made.

They plan on going to each job together and sharing the work involved. They do not anticipate hiring help.

They estimate that they will need \$10 000 for equipment, initial supplies, and their living expenses until they begin to make money. The most expensive item is a riding lawn mower.

1. List all the equipment and supplies that, in addition to the riding lawn mower, you think that Lance and Rita will need to get started.

2. Apply what you have learned about borrowing money. Which of the following ways would you recommend that Lance and Rita pay for their equipment, initial supplies, and living expenses?
   - using a credit card
   - obtaining a loan for $10 000
   - obtaining a line of credit for at least $10 000
   - obtaining overdraft protection and using their debit card and/or cheques

   Justify your recommendation.

3. Lance and Rita are preparing a business plan. They are describing the following:
   - their start-up expenses
   - the services that they will provide
   - their target market (who their customers will be)
   - how they plan to promote, or make known, their business
   - their fees
   - their expected earnings

   What factors should they take into account when determining fees for services?

4. Lance and Rita plan to provide potential customers with quotes for packages of services. If accepted, they will sign a contract. The customers will pay 10% up front and the balance after the services are provided. What might Lance and Rita do to help ensure that customers will pay the balance when it is due?

5. For single-day jobs, such as a spring yard cleanup or a hedge trimming, Lance and Rita will leave a bill due in one month when the work is finished. What should they keep in their records about each job to ensure that they are paid?

6. Other than ensuring that they know who still owes them money and for whom they still have work to do, why should Lance and Rita keep records of all the work they do for all their customers?

7. Lance and Rita plan to get contracts to remove snow from walks and driveways during the winter. They hope they will have made many contacts and provided good service to customers earlier in the year, because that should help them secure snow removal contracts. For what additional equipment will they need to plan financing?

# 9.6 Putting It All Together: Borrowing Money

In Section 8.6, you planned a trip and estimated the amount of money you would need to take it. Plan to take the trip by borrowing money for it.

1. Refer to your work from Section 8.6. How much money did you estimate that you would need?
2. List the options for borrowing money that you have learned about in this chapter. Select one to use to pay for your trip and justify your selection.
3. If you selected paying by a credit card, do the following. Use the same spreadsheet that you used in Section 9.2 to determine the cost of the trip based on current credit card interest rates and making a minimum payment of $45 each month. How much would the holiday cost you by the time you made all the payments?
4. If you selected paying by a loan (a regular loan or a line of credit), do the following. Use the same spreadsheet that you used in Section 9.4 to determine the cost of the trip based on current loan interest rates and making monthly payments for 3 years. How much would the holiday cost you by the time you repaid the loan?
5. What factors would you consider before you actually borrowed money to take a vacation?

# 9.7 Chapter Review

**1.** André is thinking about getting a credit card. He gets a phone call in which a credit card sales representative tells him he has been pre-approved for a credit card.

**a)** List five credit card features that André should ask about before he accepts the offer.

**b)** André plans to use the credit card only to defer, or put off, paying for items until the credit card due date and to always pay the amount due. What feature(s) would be most important to him?

WS **2. a)** Identify three ways other than using a credit card to borrow money from a bank for a few months or a few years.

**b)** With which of these is interest charged?

**c)** Which way has interest rates comparable to credit card interest rates?

**d)** Which way has a credit limit like a credit card?

**e)** Which way do you specify for what you want to borrow the money?

**f)** Which ways have lower interest rates than a credit card?

**g)** With which ways do banks expect security and what are two things often used as security?

AT **3.** Ryan makes a $750 purchase on his credit card. He has no previous balance and he makes no other purchases. He makes the minimum payment of $25 each month and pays 17.2% per annum compounded daily. Use the spreadsheet from Section 9.2.

**a)** Enter the values for this purchase in cells B1, B2, and B3. All the values in the columns will adjust.

**b)** Is the last new balance less than the minimum payment? If not, FILL DOWN all the columns, starting with the last row that has a value in each cell.

**c)** Select and delete the values in the rows below the first new balance that is less than the minimum payment.

**d)** Select and delete the value in the "Minimum Payment ($)" column to the right of the now last new balance. Why should you do this?

**e)** How many minimum payments are made?

**f)** How much is the final payment?

**g)** How much interest is paid altogether?

**h)** Check: Do all the minimum payments plus the final payment equal the purchase plus the total interest paid?

AT **4.** Natalie borrows money to pay for a new porch on her home.
- The amount of the loan is $10 000.
- The payments are monthly.
- The annual rate of interest is 8.5%.
- The length of the loan is 4 years.

Use the Amortization Schedule for a Loan spreadsheet from Section 9.4.

**a)** Enter the specifics for this loan in the appropriate cells.

**b)** Delete the entries, or FILL DOWN the columns so that the list goes to exactly 48 payments. The balance should now be $0.00.

**c)** How much will the new porch cost Natalie by the time she finishes repaying the loan?

**d)** Apply what you have learned about loans. Which of the following factors would have made the cost of the loan more and which would have made it less?
- borrowing more than $10 000
- making payments weekly
- having an interest rate of 7%
- repaying the loan over 5 years

AT **5.** Jonathan borrows $12 000 to buy business equipment. The annual interest rate is 6.9%. Payments are to be made monthly for 4 years.

**a)** Use the appropriate spreadsheet to determine the amount of the monthly payments and the cost of the loan.

**b)** If Jonathan doesn't feel that he can afford the monthly payments, what might he do?

CHAPTER

# 10 Buying a Car

In this chapter you will
- describe the process and cost of getting a driver's licence
- compare the process, cost, advantages, and disadvantages of buying a new car and buying a used car
- compare the cost of buying a new car and leasing the same car
- identify the factors and cost involved in insuring a car
- calculate the costs of owning and operating a car
- investigate the costs of irresponsible driving
- compare the cost of owning or leasing a car with the cost of using public transportation

Near the end of the chapter, you will determine
- the cost of buying or leasing a car that you would like to own
- the cost of insuring the car that you would like to own

# 10.1 A Driver's Licence

A driver's licence is issued by the province in which you live. It entitles you to drive a vehicle.

## Explore

What is the process and how much does it cost to get an Ontario driver's licence?

## Develop

To apply for a licence in Ontario, you must
- be at least 16 years old
- have two pieces of identification, at least one with your signature
- pay a fee
- pass a vision test
- pass a knowledge test about the rules of the road and traffic signs

To prepare for the knowledge test, you can purchase *The Official Driver's Handbook* at
- a Driver and Vehicle Licence Issuing Office
- a Driver Examination Centre
- some retail stores

All new drivers applying for a licence to drive a car (or a motorcycle) enter the **Ontario Graduated Licence System (GLS)**. This two-step process allows drivers to acquire experience and skills gradually. It takes a minimum of 20 months to complete.

Once you have met the application requirements above, you hold a **Class G1 Licence** for a minimum of 12 months or, if you have successfully completed an approved driver education course, the time is 8 months. Then, you can try the G1 Road Test.

**1.** As a Class G1 driver, you are required to do six things when operating a car. What are they? Use the Internet or *The Official Driver's Handbook* to find or verify the requirements.

Access to the Ontario Ministry of Transportation Web site can be gained through the *Mathematics for Everyday Life 11* page of irwinpublishing.com/students.

**2.** Upon passing the G1 Road Test, you hold a **Class G2 Licence** for a minimum of 12 months. Then, you can try the G2 Road Test.

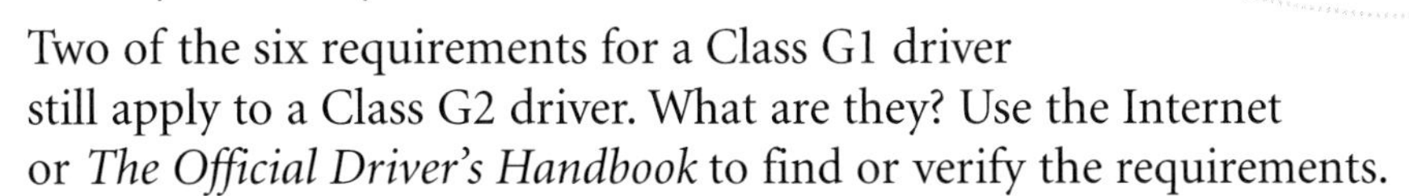

Two of the six requirements for a Class G1 driver still apply to a Class G2 driver. What are they? Use the Internet or *The Official Driver's Handbook* to find or verify the requirements.

**3.** Upon passing the G2 Road Test, you hold a full Class G Licence. How many road tests have to be passed before you can hold a Class G Licence?

## Practise

**4.** One of the requirements for a G1 driver restricts *where* you can drive.

**a)** Where are you restricted from driving?

**b)** This condition does not apply if the person with you is a driving instructor licensed in Ontario. Why would that be?

**5.** In January 2002 the licence fees were as follows.

| | |
|---|---|
| G1 Licence, including Knowledge Test, G1 Road Test, and Five-Year Licence | $100 |
| Knowledge Test | $10 |
| Five-Year Licence | $50 |
| G1 Road Test | $40 |
| G2 Road Test | $75 |

**a)** Why would the Knowledge Test, the Five-Year Licence, and the G1 Road Test be listed separately when they are all part of the G1 Licence?

**b)** What are the current fees? Use the Internet to find or verify the fees.

**6.** What is the advantage to taking an approved driver education course when getting a licence? Do you know of any other advantages once licensed? If so, what are they?

**7.** What must you have with you when you take the G1 Road Test? Use the Internet or *The Official Driver's Handbook* to find or verify what you must have.

# 10.2 A New Car or a Used Car

Buying a car is the first major purchase that many people make. Whether it is new or used, it means spending a lot of money.

## Explore

What are the advantages and disadvantages of buying a new car and of buying a used car?

## Develop

A **new car** has many costs in addition to the price on the windshield, or the **base price**, which is the price before charges, fees, and taxes. These additional costs are varying combinations of

- delivery
- freight
- pre-delivery inspection
- pre-delivery expense
- dealer preparation
- administration fee

There is a fuel consumption tax, and if the car has air conditioning, a federal air conditioner tax. PST and GST are paid on this total cost. After PST and GST, dealers often add a fuel charge because they have put some gas in the tank for you and a licence fee because they install your licence plates on your new car.

### Example 1

Juliana is thinking about buying a new subcompact car. Its total purchase price includes

$17 999 base price
$750 delivery charge
$100 federal air conditioner tax
$75 fuel consumption tax
8% PST on the above
7% GST on the same as the PST
$20 licence fee
$25 fuel

There is a warranty for three years or 60 000 km.

**a)** What will the total purchase price of this new subcompact car be?

**b)** How much more is the total purchase price than the base price?

### Solution 

**a)** price before PST and GST = base price + delivery charge + air tax + fuel tax
= 17 999 + 750 + 100 + 75
= 18 924

To find the price including PST and GST, you can calculate 8% of 18 924, 7% of 18 924, and then add both to 18 924.

Or, you can find 115% of 18 924 because the price before PST and GST is 100%, PST is 8%, and GST is 7%.

price including PST and GST = 115% of price before PST and GST
= $1.15 \times 18\,924$
= 21 762.60

total purchase price = price including PST and GST + licence fee + fuel
= 21 762.60 + 20 + 25
= 21 807.60

The total purchase price of this new subcompact car will be $21 807.60.

**b)** total purchase price − base price = 21 807.60 − 17 999
= 3808.60

The total purchase price is $3808.60 more than the base price.

A new car **depreciates**, or loses value, quickly. The amount of depreciation depends on the model of car and how many kilometres it is driven. A typical car made by a North American manufacturer depreciates about 50% in three years.

## Example 2

Assume that the car that Juliana, from Example 1, is thinking about buying will depreciate 40% in two years. How much will the car be worth in two years?

### Solution 

Depreciation is calculated on the base price. The car will be worth 40% less than the base price, or 60% of the base price.

value after two years = 60% of base price
= $0.60 \times 17\,999$
= 10 799.40

The value of the car will be $10 799.40 after two years.

Before a **used car** can be sold, it must

- receive a **Safety Standards Certificate**, indicating that it has been inspected and repaired (if necessary), and is safe to drive
- pass a **Drive Clean emissions test** (where applicable), indicating that it emits, or gives off, an allowable amount of pollutants

The applicable fees may or may not be included in the price of a used car.

If you buy a used car from a dealer, you pay both PST and GST. After PST and GST, dealers often add a fuel charge because they have put some gas in the tank for you and a licence fee because they install your licence plates on your car. Dealers may provide a three-month warranty and offer to sell you an additional warranty.

If you buy a used car privately, you pay only one tax, PST, directly to the government.

## Example 3

Juliana, from the previous examples, is also thinking about buying a two-year-old used car from a dealer.

- The price, including the Safety Standards Certificate and the Drive Clean emissions test, is $11 500.
- The licence fee and the fuel are each $20.
- There is a three-month warranty, and she could buy an additional warranty for one year or 25 000 km for $1200 plus PST and GST.

What is the total purchase price of the car with the additional warranty?

### Solution 

price before taxes = car + additional warranty
= 11 500 + 1200
= 12 700

price including PST and GST = 115% of price before taxes
= 1.15 × 12 700
= 14 605

total purchase price = price including PST and GST + licence fee + fuel
= 14 605 + 20 + 20
= 14 645

The total purchase price with the additional warranty is $14 645.

## Practise

1. Consider the information in the preceding examples. Discuss the advantages and disadvantages of buying new and buying used. If you were Juliana, which car buying option would you choose? Justify your choice.

2. Gopal is thinking about buying a new compact car. Its total purchase price includes

   $25 400 base price
   $690 pre-delivery expense
   $300 freight
   $100 federal air conditioner tax
   $75 fuel consumption tax
   $98 administration fee
   8% PST on the above
   7% GST on the same as the PST
   $20 licence fee
   $30 gas

   The car comes with a warranty for three years or 60 000 km.

   **a)** What is the total purchase price?

   **b)** Assume that the car will depreciate 30% in the first year. How much will the car be worth in a year?

3. Gopal finds the previous year's model of the car in question 2 at a used car dealer. Its total purchase price includes

   $20 000 price, including the Safety Standards Certificate and the Drive Clean emissions test
   8% PST on the above
   7% GST on the same as the PST
   $20 licence fee
   $25 fuel

   The car comes with a three-month warranty, and he can buy an additional warranty for one year or 25 000 km for $1250 plus PST and GST.

   **a)** What is the total purchase price without the additional warranty?

   **b)** What is the total purchase price with the additional warranty?

4. Gopal also finds the same one-year-old car as in question 3 advertised for sale privately for $20 000.

   - The Safety Standards Certificate will cost $60 and the Drive Clean emissions test will cost $30.
   - The owner expects Gopal to pay for these, including PST and GST on both fees, but if any work is required, she will pay for it.
   - Gopal must pay 8% PST on the price of the car.
   - There are no warranties available.

   **a)** What is the total purchase price?

   **b)** What questions should Gopal ask the owner before he decides to buy the car?

**5.** For someone like Gopal wanting to buy a new or almost new compact car, discuss the advantages and disadvantages of each car buying option from questions 2 to 4.

**6.** Determine the depreciated value of each car.

**a)** base price $18 500, depreciated 25% in one year
**b)** base price $29 400, depreciated 35% in two years
**c)** base price $20 600, depreciated 45% in three years
**d)** base price $24 800, depreciated 50% in three years

**7.** When you buy a car, you will also have to pay for
• a vehicle permit and licence plates
• insurance

You can't drive your car without these items and a driver's licence. You investigated the process and the cost of getting a driver's licence in Section 10.1. In Section 10.5, you will investigate the cost of insurance.

Use the Internet to find the fees for

**a)** a vehicle permit and number plates (standard licence plates) for a car
**b)** a yearly vehicle validation (licence plate sticker) for a car

Access to the Ontario Ministry of Transportation Web site can be gained through the *Mathematics for Everyday Life 11* page of irwinpublishing.com/students.

**8.** Identify the following for buying new, buying used from a dealer, and buying used privately.

Price, fees, and so on involved before sales taxes
Sales taxes that apply
Fees and so on after taxes
Warranties available
Advantages
Disadvantages

## Skills Check Operations with Decimals

Calculate.

**a)** $2500 + 48 × $289.50
**b)** $3000 + 60 × $472.80
**c)** $3000 + 36 × $612.60
**d)** $2000 + 60 × $503.75
**e)** $4000 + 48 × $315.40
**f)** $3500 + 48 × $365.10
**g)** $2800 + 48 × 115% × $379
**h)** $3200 + 48 × 115% × $344

# 10.3 Buying Versus Leasing

Only 15% of Canadians pay cash for a new car. In Chapter 9, you investigated the cost of borrowing money for purchases such as a car.

## Explore

If you want a new car, an alternative to buying is leasing. What does leasing mean? How is leasing different from buying? How is it the same?

## Develop

**1.** Sung would like to either buy or lease a new car with a total purchase price of \$26 600. He has \$3000 for a down payment. If he buys, he can get a loan for the remaining amount. It would be repaid in 48 monthly payments of \$576.14.

**a)** In how many years will he own the car?

**b)** How much will he pay altogether for the car?

**c)** How much more will he pay than if he paid cash?

### Example

Sung can **lease**, or rent, the car in question 1 for

- \$3000 down payment
- \$300 refundable security deposit
- 48 monthly payments of \$324.24
- PST and GST on each monthly payment

The monthly payments would be more manageable than the loan payments. When the lease is up, Sung can buy the car for \$12 000 plus PST and GST.

**a)** How much will he pay to use the car for four years?

**b)** If he buys the car when the lease is up, how much will he pay altogether?

### Solution

**a)** lease payment including taxes $= 115\%$ of monthly payment  100% payment + 8% PST + 7% GST

$= 1.15 \times 324.24$

$= 372.88$

cost of the lease $=$ down payment $+ 48 \times$ lease payment including taxes

$= 3000 + 48 \times 372.88$

$= 20\ 898.24$

Assuming he gets his \$300 security deposit back, Sung will pay \$20 898.24 to use the car for four years.

**b)** price after lease is up $= 115\%$ of $\$12\,000$   100% price + 8% PST + 7% GST

$= 1.15 \times 12\,000$

$= 13\,800$

total cost $=$ cost of lease $+$ price after lease is up

$= 20\,898.24 + 13\,800$

$= 34\,698.24$

If he buys the car after the lease is up, Sung will pay $34 698.24 altogether.

**2.** Because you don't own a leased car, you must pay for damages, such as scratches, dents, chipped windshield, and bald tires, when you return it. As well, there is a charge if you drive more than the number of kilometres specified in the agreement.

**a)** Discuss the advantages and disadvantages of buying a new car with a loan and of leasing a new car.

**b)** If you were Sung, would you buy or lease? Explain your decision.

## Practise

**3.** Calculate the monthly lease payment including PST and GST for each of the following monthly payments which are before taxes.

**a)** $380 **b)** $452 **c)** $289 **d)** $402.60

**4.** Calculate the total amount to be paid before driving the leased car off the lot for each situation.

**a)** $1500 down payment, $250 refundable security deposit, $209 first monthly payment, PST and GST on first monthly payment

**b)** $2000 down payment, $300 refundable security deposit, $278 first monthly payment, PST and GST on first monthly payment

**c)** $3500 down payment, $350 refundable security deposit, $418.50 first monthly payment, PST and GST on first monthly payment

**5.** Manjusha is going to buy or lease a new car with a base price of $17 500 and a total purchase price of $21 520. She has $2500 for a down payment. She can get a loan for the remaining amount and pay 60 monthly payments of $358. Or, she can make the same down payment and lease for 48 months. The lease payments would be $228 plus PST and GST. When the lease is up, Manjusha can buy the car for $8500 plus PST and GST.

**a)** If she buys, in how many years will she own the car?

**b)** If she buys, how much will she pay altogether for the car?

**c)** If she buys, how much more will she pay than if she paid cash?

**d)** If she leases, how much will she pay to use the car for four years?

**e)** If she buys the car when the lease is up, how much will she pay altogether?

**f)** If you were Manjusha, would you buy or lease the car? Explain your decision.

# 10.4 Career Focus: Car Salesperson

Maureen sells new cars. When she was in high school, she took marketing courses. As well, she has taken a course from the Canadian Automotive Institute to help her in her chosen career. Maureen is outgoing and likes to meet people. She enjoys helping people select cars that suit their needs. She has good math skills for calculating and explaining the many additional costs associated with new cars.

**1.** Ted and Lise want to buy the car described below from Maureen. The total purchase price includes

   $18 090 base price
   $795 freight
   $650 dealer preparation
   price of each option selected
   $100 federal air conditioner tax, if applicable
   $75 fuel consumption tax
   8% PST on the above
   7% GST on the same as the PST

**Compact 4-door sedan**

*Standard Equipment*

| | | |
|---|---|---|
| electric trunk release | console | anti-lock brakes |
| roadside assistance program | fixed mast antenna | folding rear back seat |
| tinted glass | hi, low, and fixed delay wipers | front disc, rear drum brakes |
| steel wheels | daytime running lights | stainless steel exhaust system |
| rear window defogger | front and rear mats | 2.2 litre engine |
| AM-FM stereo with front and rear speakers | reclining bucket front seats | 5-speed manual transmission |
| | all-season radial black wall tires | |

*Optional Equipment*

| | | | |
|---|---|---|---|
| 4-speed automatic transmission | $1180 | special edition side moldings and aluminum wheels | $1050 |
| air conditioning | $1150 | sunroof $805 | |
| cruise control | $365 | aluminum wheels | $375 |
| AM-FM stereo with 6 speakers | $415 | rear spoiler | $235 |
| premium audio system with 8 speakers | $265 | chrome wheel cover | $75 |
| licence plate cover | $15 | | |

Ted and Lise selected the following options.
- 4-speed automatic transmission
- air conditioning
- AM-FM stereo with 6 speakers
- aluminum wheels
- rear spoiler

What is the total purchase price that Maureen calculates?

**2.** Suppose you wanted to buy the same car. Which options would you select? What is the total purchase price that Maureen would calculate?

3. Most car salespersons are paid a commission of 25% to 40% of the dealer's profit, or the difference between the purchase price and the dealer's cost. Maureen earns 30% commission. If the dealer's profit on Ted and Lise's car is $1350, what does Maureen earn making that deal?

4. Maureen also has the following bonus plan.

| Number of Cars Sold per Month | Bonus |
|---|---|
| 7 to 9 | $500 |
| 10 to 12 | $800 |
| 13 to 15 | $1200 |
| 16 or more | $1800 |

Her car sales and leases for one year are shown in the spreadsheet below.

a) Explain how the amounts given in cells D2, E2, and F2 were calculated.

b) FILL DOWN Columns D to F, Rows 2 to 13. What are Maureen's gross monthly earning for each of the other months?

c) In which months were her gross monthly earnings the same as the commission she earned? Why?

| | A | B | C | D | E | F |
|---|---|---|---|---|---|---|
| 1 | Month | Number of Maureen's Sales or Lease Deals | Dealer's Profit on Maureen's Deals ($) | Commission Earned ($) | Bonus Earned ($) | Maureen's Gross Monthly Income ($) |
| 2 | Jan. | 6 | 9600 | 2880.00 | 0.00 | 2880.00 |
| 3 | Feb. | 5 | 7465 | | | |
| 4 | Mar. | 8 | 13768 | | | |
| 5 | Apr. | 10 | 12470 | | | |
| 6 | May | 12 | 25572 | | | |
| 7 | June | 15 | 27520 | | | |
| 8 | July | 7 | 8057 | | | |
| 9 | Aug. | 5 | 6340 | | | |
| 10 | Sept. | 9 | 14589 | | | |
| 11 | Oct. | 11 | 13857 | | | |
| 12 | Nov. | 8 | 20240 | | | |
| 13 | Dec. | 3 | 3490 | | | |

# 10.5 Insuring a Car

Owning and operating a car has many risks. Even the best drivers can be involved in accidents through no fault of their own. By law, the owner of a car is required to have insurance on the car. Failure to have insurance results in a hefty fine and a court appearance.

## Explore

Janice's car insurance **premium**, or what she pays for insurance coverage, is $1450 a year. Wayne's car insurance premium is $3800 a year. Identify some of the factors that affect insurance premiums.

## Develop

Car insurance premiums vary according to the likelihood of you being involved in a collision or having your car stolen. The insurance industry keeps detailed statistics about **claims**, or requests for insurance companies to pay, in order to determine the likelihood.

There are two aspects to insurance coverage:

- to compensate (or pay) others for their losses
- to pay for expenses related to your car

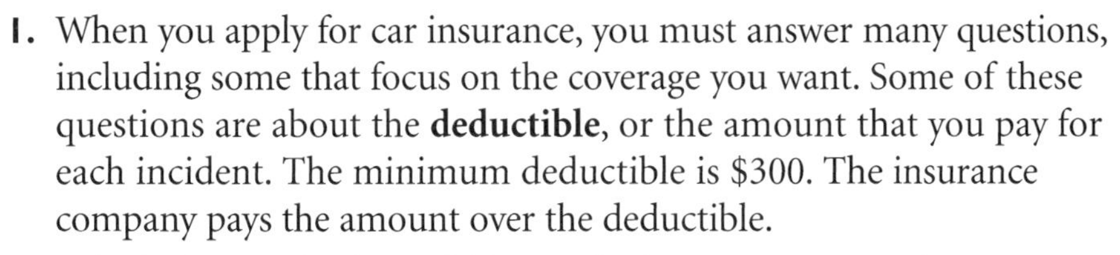

1. When you apply for car insurance, you must answer many questions, including some that focus on the coverage you want. Some of these questions are about the **deductible**, or the amount that you pay for each incident. The minimum deductible is $300. The insurance company pays the amount over the deductible.

   Which do you think would have a higher premium—a $300 deductible or a $600 deductible? Why?

2. Other questions focus on the car that you are insuring. In addition to cars varying greatly in value, some models of cars are stolen more often, some are more expensive to repair, and some offer passengers more protection.

   Give an example of a car that would cost more to insure than a two-year-old subcompact.

**3.** Most of the questions when applying for car insurance focus on you—your driver's licence, your driving history, and your use of the car.

Which, in each pair of situations, do you think would have a higher premium? Why?

**a)** being 18 years old or being 36 years old
**b)** living in Toronto or living in the rural Ottawa Valley
**c)** getting a full Class G Licence last year or getting it 25 years ago
**d)** completing an approved driver education course or not taking such a course
**e)** having no tickets in the last six years or having four tickets in the last six years
**f)** never having had your driver's licence suspended or having had your driver's licence suspended
**g)** having had a car insurance policy cancelled or never having a car insurance policy cancelled
**h)** driving the car less than 10 000 km a year or driving the car more than 30 000 km a year

## Practise

**4.** Gord just received a Class G Licence. He must be insured to drive his parents' car. Their premium is $1860 a year. Adding Gord as a driver will cost 60% more. Gord can use his parents' car if he pays the difference in the premium. How much will Gord have to pay?

**5.** Age, gender, and marital status are questions that cannot be asked when applying for a job. However, they can be, and are, asked when applying for car insurance. Why might that be?

**6.** If you are licensed to drive, obtain quotes for an insurance premium. Or, if you are not licensed to drive, obtain quotes for someone who is. Obtain the quotes for a car that you would like to have.

Access to a Web site for obtaining car insurance quotes can be gained through the *Mathematics for Everyday Life 11* page of irwinpublishing.com/students.

**7.** Justify the following statement.

If you are thinking about buying a car, you should consider the cost of insurance as part of what you must be able to afford.

# 10.6 Owning and Operating Costs

## Explore

Owning and operating a car has many costs. You have seen some of these costs. What are they? What are the other costs?

## Develop

Owning and operating costs can be calculated annually, monthly, and per kilometre. Several costs are given as annual amounts. Car payments, if applicable, are monthly (or weekly or biweekly or semi-monthly). Amounts paid for other costs must be recorded and totalled. As well, the distance driven in a year needs to be calculated.

### Example 1

An automobile association calculates the annual cost of a specific small car to be $10 512. Among the factors taken into account to determine the cost are car payments and depreciation. Assuming that 24 000 km are driven in a year, calculate the monthly cost and the cost per kilometre.

### Solution

$$\begin{aligned}\text{monthly cost} &= \text{annual cost} \div 12\\ &= 10\,512 \div 12\\ &= 876\end{aligned}$$

$$\begin{aligned}\text{cost per kilometre} &= \text{annual cost} \div \text{kilometres driven per year}\\ &= 10\,512 \div 24\,000\\ &= 0.438\end{aligned}$$

The cost of owning and operating that car is $876 per month or $0.438/km or 43.8¢/km.

## Practise

1. You need a driver's licence and your car needs licence plates and, each year, a new licence plate sticker. In Sections 10.1 and 10.2, you found these costs. Calculate the cost to get a driver's licence, standard plates, and a sticker

   **a)** where you live **b)** in the other part of Ontario

2. The cost of gasoline is a major variable cost. How much does it cost to fill up a tank that holds 60 L when gasoline sells for 72.3¢/L?

3. Maintenance and repair costs vary with the model of car and usually increase as the car gets older. Alison's car needs four new tires. She buys four for $95 each. Installation and balancing cost $20 per tire. What is the total cost of four tires before PST and GST?

4. Last year the annual cost of George's car was $11 318. He drove the car 21 085 km. Calculate the monthly cost and the cost per kilometre.

5. Last year the cost per kilometre of Tanya's car was 45.2¢/km. She drove the car 19 038 km. Calculate the annual cost and the monthly cost.

6. **a)** Explain how the amounts given in cells J2, K2, and L2 were calculated.

   AT **b)** FILL DOWN Columns J to L, Rows 2 to 8. What are the annual costs, monthly costs, and the costs per kilometre for each of the other cars?

| | A | B | C | D | E | F | G | H | I | J | K | L |
|---|---|---|---|---|---|---|---|---|---|---|---|---|
| 1 | Car | Distance Driven (km) | Gasoline ($) | Oil ($) | Maintenance/ Repairs ($) | Insurance ($) | Licence Sticker ($) | Depreciation ($) | Monthly Car Payments ($) | Annual Cost ($) | Monthly Cost ($) | Cost per Kilometre ($) |
| 2 | Damien's | 21380 | 1800 | 114 | 1024 | 1250 | 74 | 1400 | 268 | 8878 | 739.83 | 0.42 |
| 3 | Clare's | 16560 | 1325 | 92 | 316 | 1870 | 37 | 1850 | 0 | | | |
| 4 | Toma's | 14990 | 1120 | 85 | 85 | 2680 | 74 | 1480 | 298 | | | |
| 5 | Gilda's | 26710 | 1960 | 88 | 140 | 1490 | 37 | 2890 | 345 | | | |
| 6 | Cale's | 18340 | 1460 | 92 | 105 | 2350 | 37 | 1910 | 312 | | | |
| 7 | Moufeed's | 19030 | 1530 | 88 | 0 | 1690 | 74 | 2440 | 0 | | | |
| 8 | Janet's | 17670 | 1380 | 92 | 220 | 2160 | 74 | 1990 | 345 | | | |

7. All of the costs in question 6 vary. The costs of gasoline and oil, for example, vary depending upon the car being driven, the distance driven, the driving conditions, and the cost per litre.

   Why do each of the other costs vary?

# 10.7 The Costs of Irresponsible Driving

In order to keep the roads safe, it is important that everyone obey the laws of the roads and highways. Driving is a privilege, not a right.

## Explore

What is a demerit point? Other than paying a fine for speeding, how might a speeding conviction cost you money?

## Develop

**1.** Use the Internet or *The Official Driver's Handbook* to find out more about the **demerit point system**, and answer the following questions.

**a)** How many demerit points do you receive for following too closely?

**b)** For what will you receive seven demerit points?

**c)** How long do demerit points stay on your driving record?

**d)** At how many points is a Class G1 or G2 Licence suspended?

**e)** At how many points is a Class G Licence suspended?

Access to the Ontario Ministry of Transportation Web site can be gained through the *Mathematics for Everyday Life 11* page of irwinpublishing.com/students.

## Practise

**2.** The following penalties for speeding were effective January 1, 2002.

| Speed over the Speed Limit (km/h) | Amount of Fine | Number of Demerit Points |
|---|---|---|
| 15 | $42.50 | 0 |
| 20 | $80.00 | 3 |
| 25 | $98.75 | 3 |
| 30 | $117.50 | 4 |
| 35 | $215.00 | 4 |
| 40 | $245.00 | 4 |
| 45 | $275.00 | 4 |
| 49 | $299.00 | 4 |

If charged with going 85 km/h in a 60 km/h zone, what fine must you pay? How many demerit points would you receive?

3. Use the speeding penalties table on page 199 and the surcharge table below.

The following surcharges, or extra fines, were effective January 1, 2002.

| Amount of Fine | Surcharge |
|---|---|
| $0 – $50 | $10 |
| $51 – $75 | $15 |
| $76 – $100 | $20 |
| $101 – $150 | $25 |
| $151 – $200 | $35 |
| $201 – $250 | $50 |
| $251 – $300 | $60 |
| and so on | |

a) What is the total of the fine and surcharge for exceeding the speed limit by 20 km/h?

b) What is the total of the fine and surcharge for going 135 km/h in a 100 km/h zone?

4. If charged with going over the speed limit by 50 km/h or more, you must appear in court. What costs other than paying the fine and the surcharge might you have?

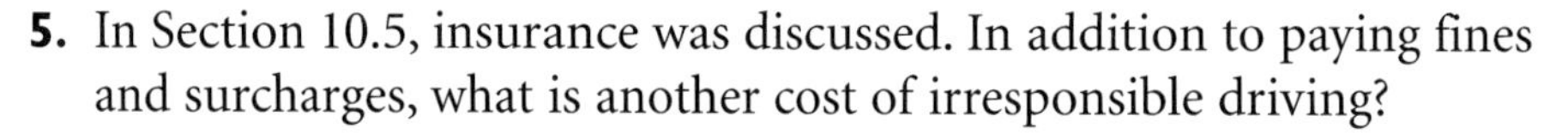

5. In Section 10.5, insurance was discussed. In addition to paying fines and surcharges, what is another cost of irresponsible driving?

6. Some areas are designated Community Safety Zones. If you were charged with speeding in a Community Safety Zone, the fine and surcharge could both be double. What might you pay for exceeding the speed limit by 20 km/h in a Community Safety Zone?

7. The penalty for not wearing a seatbelt ranges from $90 to $500 and two demerit points. The penalty for failing to stop for a school bus ranges from $400 to $2000 and six demerit points. If you were a Class G1 or G2 driver and were charged with both offences, how many points away from having your licence suspended would you be?

# 10.8 A Car Versus Public Transportation

## Explore

What means of transportation other than owning a car are available to people in your community? Which would cost less than owning and operating a car? Which would cost more? Explain.

## Develop

**Use the following information for questions 1 to 5.**

Trent lives in Thunder Bay. He needs transportation to and from work and for recreational use. He will drive about 20 000 km a year.

1. Trent might buy a new car for a total purchase price of \$20 760. It is estimated that the cost, which includes car payments and depreciation, will be 40¢/km. Calculate Trent's annual and monthly costs.

2. Trent could lease the same car as in question 1. It is estimated that the cost, which includes lease payments but not depreciation because he does not own the car, will be 28¢/km. Calculate Trent's annual and monthly costs.

3. Trent finds a used car for a total purchase price of \$12 500. The car payments and depreciation are less, but the maintenance and repairs will be more. It is estimated that the cost will be 37¢/km. Calculate Trent's annual and monthly costs.

4. Trent could take public transportation to and from work. A monthly transit pass costs \$58. For local recreation, he could use a combination of public transportation and taxis. He estimates that he might spend \$250 a year for taxis. To go out of town, he could rent a car, take a train, or take a bus. He might spend \$500 on out-of-town transportation. What does Trent estimate that he will spend on transportation if he doesn't get a car?

5. Trent does not have to pay for parking where he works.
   a) Rank the transportation options from questions 1 to 4 from most to least expensive.
   b) What factors other than cost should he take into account before deciding?
   c) If you were Trent, which option would you choose? Why?
   d) Discuss the advantages and disadvantages of having a car.
   e) Discuss the advantages and disadvantages of not having a car.

## Practise

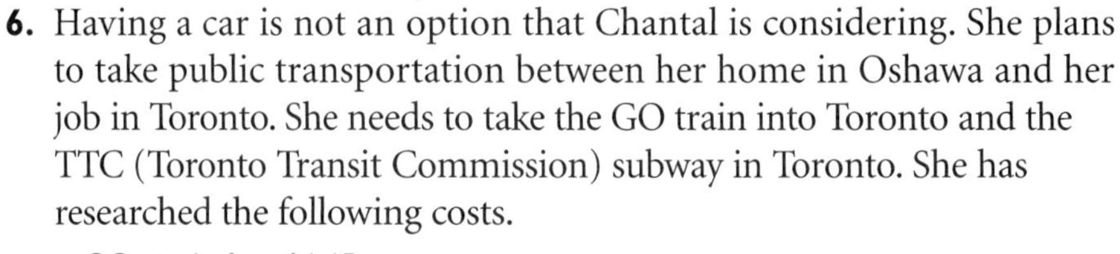

6. Having a car is not an option that Chantal is considering. She plans to take public transportation between her home in Oshawa and her job in Toronto. She needs to take the GO train into Toronto and the TTC (Toronto Transit Commission) subway in Toronto. She has researched the following costs.

   GO single fare $6.65
   GO monthly pass $210
   TTC single fare $2.25
   TTC monthly pass $93.50
   GO/TTC monthly Twin Pass $293.50

   Assume that she will commute to and from work 20 days a month.
   a) Which is the most economical option?
   b) How much less than the most expensive is the most economical?
   c) Chantal takes two weeks of vacation in August. What is the most economical option for getting to and from work that month?
   d) Discuss the advantages and disadvantages of using public transportation.

7. Shair and Jeremy don't know each other. They live in different areas of Ottawa, but work in the same area. They are both deciding whether or not to get a car. Parking in the area where they work is $8 a day and a monthly public transit pass costs $73.50. They investigate and find out a monthly parking pass is available for $120.

   Shair immediately dismisses any thoughts of getting a car. Jeremy, however, is going to investigate the cost of a car. Why might each have made the decision he did?

8. Research the costs of public transportation in your community. If you live someplace where people often commute to another community to work, research transportation costs between communities. How do those costs compare with the costs for cars that you have already seen?

# 10.9 Putting It All Together: Buying a Car

**A** Complete the following research project and identify your sources of information.

- Identify a new or used car that you would like to own.
- Determine the total purchase price to buy the car and the terms of a lease if leasing is an option.
- Obtain quotes for the insurance premium that you would have to pay to drive the car. (If you are not licensed to drive, obtain quotes for a person who has had a Class G Licence for a year, but never had a car or insurance before.)

# 10.10 Chapter Review

**1.** **a)** What is the minimum amount of time needed to complete the Ontario Graduated Licence System?

**b)** What licences must you hold before getting a full Class G Licence?

**c)** How many road tests must you pass to get a full Class G Licence?

**d)** What are three of the six things that you must do when operating a car with a first stage licence?

**e)** How often must a full Class G Licence be renewed? What is the current fee?

**2.** Enida is thinking about buying a new subcompact car. Its total purchase price includes

$16 800 base price
$600 pre-delivery expense
$300 freight
$100 federal air conditioner tax
$75 fuel consumption tax
$98 administration fee
8% PST on the above
7% GST on the same as the PST
$20 licence fee
$25 gas

The car comes with a warranty for three years or 60 000 km.

**a)** What is the total purchase price?

**b)** Assume that the car will depreciate 25% in the first year. How much will the car be worth in a year?

**3.** Bryan sees a one-year-old subcompact at a used car dealer. Its total purchase price includes

$15 990 price, including the Safety Standards Certificate and the Drive Clean emissions test
8% PST on the above
7% GST on the same as the PST
$20 licence fee
$25 fuel

The car comes with a three-month warranty, and Bryan can buy an additional warranty for one year or 25 000 km for $1200 plus PST and GST.

What is the total purchase price with the additional warranty?

**4.** To lease a car, a down payment of $3000, a refundable security deposit of $300, the first monthly payment of $299, and PST and GST on the first monthly payment must be paid. Calculate the total amount to be paid before driving the car off the lot.

5. Discuss six factors that affect car insurance premiums. Include whether each factor increases or decreases the premium.

6. Celine just received a Class G Licence. She must be insured to drive her parents' car. Their premium is $1650 a year. Adding Celine as a driver will cost 35% more. Celine can use her parents' car if she pays the difference in the premium. How much will Celine have to pay?

7. The annual costs for Liz's new car the first year were

   $1250 gasoline
   $108 oil
   $130 maintenance and repairs
   $1420 insurance
   $74 licence sticker
   $5200 depreciation
   $425 monthly car payments

   She drove the car 18 420 km. Determine each cost.

   **a)** annual **b)** monthly **c)** per kilometre

8. In what ways can failing to obey the rules of the road cost you money? What are other possible consequences?

9. Maria lives in your community. She does not live close enough to her place of work to walk. What public transportation options are available to her? What are the fares?

10. Aaron is single. He has a job with a weekly take-home pay of $482. His monthly expenses, including a $55 bus pass, are $1185. He has been saving money for three years and has $3000 to make a down payment on a car. Identify all the costs associated with owning and operating a car. Discuss the advantages and disadvantages of buying a new car, leasing a new car, buying a used car, and using public transportation. Apply what you have learned to help Aaron make a wise decision about buying a car.

# Glossary

## A

**ABM (Automated Banking Machine)** a machine, found in such places as banks, malls, variety stores, and grocery stores, which allows the use of a bank card to withdraw money

**accommodation** lodgings, such as an inn, hotel, motel, bed and breakfast, and campground

**administration fee** a charge for special services, such as delaying payment

**amortization schedule** a tabular arrangement by payment date, interest paid, principal paid, and balance at the time of each payment on a loan

**amount (of an investment)** the principal that was invested plus the interest earned

**annual** yearly

**annum** a year

**arrival time** time at which one reaches a destination

**ATM (Automated Teller Machine)** a machine, found in such places as banks, malls, variety stores, and grocery stores, which allows the use of a bank card to withdraw money

**automated banking** self-service banking, or banking done at a bank machine, over the Internet, or by telephone

**automatic withdrawal** a regular, prearranged withdrawal from an account

## B

**bank statement** a monthly record showing the transactions and balance in a bank account

**base price (of a new car)** the sticker price, or the price before assorted charges, fees, and taxes

**benefits** services paid for in full or in part by an employer, for example, dental coverage, life insurance, and health insurance

**blue-chip stocks** shares in large, well-established corporations

**bond** a written promise by a government or company to repay borrowed money, usually with interest

**bonded** being insured against stealing

**bonus** a single payment for good employee or company performance

## C

**Canada Savings Bonds** bonds that are issued by the Government of Canada; these pay simple or compound interest and are cashable at any time.

**capital gain** an increase in the value of an investment, excluding interest or dividends

**capital loss** a decrease in the value of an investment, excluding interest or dividends

**CCRA (Canada Customs and Revenue Agency)** the federal government department that administers tax laws

**cheque** a slip of paper that provides written instructions to a bank to pay an amount to the person or organization named

**chequing account** a bank account that permits the writing of cheques; the type used by most people for receiving pay and paying expenses

**cigarette taxes** the portion of the selling price of cigarettes that consists of federal and provincial government taxes other than GST and PST

**claim** a request for an insurance company to pay for damages or loss

**Class G Licence** the driver's licence for passenger vehicles in Ontario, obtained after completing the Class G1 and G2 Road Tests

**Class G1 Licence** the first stage for new drivers of passenger vehicles in Ontario in the process of earning full driving privileges

**Class G2 Licence** the second stage for new drivers of passenger vehicles in Ontario in the process of earning full driving privileges

**commission** money earned for selling a product or service, based on a percent of the value of the items sold or the number of items sold where the percent changes

**compound interest** interest calculated on the principal and the accumulated interest

**compounding period** the time for which compound interest is calculated, such as monthly or semi-annually

**cost of a loan** the total interest paid upon full repayment of a loan

**CPP (Canada Pension Plan)** a fund operated by the federal government to which workers between 18 and 70 and their employers contribute and from which payments to workers are made when they retire

**credit card** a card authorizing purchases to be made but paid for later

**credit limit** the maximum balance owing that is permitted on a credit card, line of credit, or other type of loan

**currency** money in circulation

**customs** a tax on goods

## D

**debit card** a bank card used to pay for purchases where the money is transferred by computer from the purchaser's bank account to the store's bank account

**deductible** the amount the insured pays for an insured incident, such as a car accident, and the amount over which the insurance company pays

**deductions** an amount taken off the gross pay for such items as taxes, pension plans, health plans, employment insurance, union dues, and payroll savings plans

**deferral charge** a fee for paying for an item at a later date

**demerit point system** where drivers convicted of certain driving-related offences have points recorded against their records for two years; if there are 15 points, the licence is suspended.

**departure time** time at which one leaves

**deposit** put money in a bank account

**depreciation** the decrease in the value of property, such as a car, because of age and wear

**destination** a location set as the end of a journey
**direct deposit** money transferred by computer to one bank account from another account
**direct tax** when the exact amount of the tax is evident, as opposed to being included in the price
**discount** the difference between the regular selling price and the sale price
**discount coupon** an ad of a sale price, distributed by a store or manufacturer, where the ad must be handed in to get the sale price
**dividend** a sum payable as profit to the holders of stock in a company
**double time** an overtime hourly rate equal to two times the regular hourly rate
**Drive Clean emissions test** a test to determine the level of pollutants that a vehicle in Ontario emits; failure means that the vehicle must be repaired or taken off the road.
**duty** a tax on goods

## E

**early buyout** the opportunity to pay off the amount owed early, often with an additional fee charged
**earnings** money paid for work done
**EI (Employment Insurance)** a fund operated by the federal government to which employees and their employers contribute; if employees lose their jobs, through no fault of their own, they may receive monthly payments from the fund for a specified period of time.
**electronic funds transfer** the movement of money from one account to another using computers
**employee** a person who works for, and is paid by, another person or a company
**employer** the person or company that pays workers for work done
**entry-level position** a job that requires minimal skill or training
**essential expenses** things that must be paid for in order to live, such as housing, food, clothing, and health care
**estimate** a rough calculation to achieve an approximate value; often made using rounded numbers
**exchange rate** the value of one currency in terms of another

## F

**face value (of a bond)** stated value of a bond
**frequency (of pay)** how often one is paid, such as weekly and monthly
**full time** employment that involves usually 35 hours of work a week and full benefits
**full-service banking** banking done in the bank with the help of a teller

## G

**gasoline taxes** the portion of the selling price of gasoline that consists of federal and provincial government taxes excluding GST
**General Income Tax and Benefit Guide** a booklet provided by the Canada Customs and Revenue Agency with instructions about completing an income tax return
**GIC (Guaranteed Investment Certificate)** an investment where the interest rate remains the same for a set time
**gross income** money earned before deductions
**gross pay** money earned before deductions
**GST (Goods and Services Tax)** a federal government tax on most goods and services, 7% of the price in 2002

## H

**health plan** a plan which partially or fully covers medical costs not covered by OHIP (Ontario Health Insurance Plan)
**hourly rate** the amount a worker is paid per hour

## I

**incentive (to purchase)** a feature that motivates one to buy
**income return** interest from bonds or dividend from stocks
**income tax** a federal and provincial government tax on earnings
**income tax return** an official form completed to determine how much income tax should be paid for a year; it is sent to the Canada Customs and Revenue Agency along with any tax due.
**indirect tax** a tax included in the retail price of goods, such as gasoline and cigarette taxes
**information slip** an official record of income earned or interest income
**instalment plan** a payment contract that involves making payments at set intervals for an item purchased
**interest** a fee charged/paid for borrowing/loaning money, usually a percent of the amount of money involved
**invest** to set money aside and put it to work to earn more money over the long term
**investment objective** the goal of an investment plan—safety of principal, income, or growth
**itinerary** the route of a journey or the planned outline of one

## J

**jewellery tax** a federal government tax on jewellery

## L

**layaway** having a store hold a purchase until a later date by making a payment to indicate intent to buy later
**lease** agreeing to pay a set fee at regular intervals for a given time for use of property, such as a car or a house
**licensed** possessing the legal documents to do something specific
**line of credit** a fixed, pre-authorized amount of money that is available to borrow with interest charged
**list price** the suggested retail price of an item
**loan** money to be used, but paid back with interest

## M

**market value (of a bond)** current value of a bond
**markup** the difference between the prices the retailer paid for items and the prices the retailer charges the customer
**mature** come due or reach the end of a term
**merchandise** anything bought and sold
**minimum payment** the smallest amount that must be paid on the balance of credit owing upon receipt of a statement of account

**minimum wage** the lowest hourly rate of pay allowed by law
**monthly fee** a charge for services that is paid each month
**mortgage** a loan secured by property
**mutual fund** a large collection of stock and/or bonds owned by many people, who purchase units, and administered by a fund manager

## N

**net income** money earned less deductions
**net pay** money earned less deductions
**new balance** amount owing on a credit account
**non-essential expenses** things that are paid for but are not necessary to live, for example, entertainment and a cell phone

## O

**on-line banking** banking done over the Internet
**Ontario Graduated Licence System** the two-step licensing process where new drivers obtain experience and skills before they gain full driving privileges
**overdraft protection** credit extended in case of withdrawing more money from a bank account than there is in the account
**overtime** time worked beyond regular hours
**overtime rate** the hourly rate paid for hours worked beyond regular hours

## P

**part time** employment that involves fewer hours than full time and fewer or no benefits
**passbook** a small book in which banking transactions and balances are recorded
**payroll savings plan** an arrangement by which money is deducted from gross income and invested on behalf of the employee
**peak season** time of year at which attendance or participation is at a maximum
**pension plan** money deducted from gross income and invested on behalf of the employee for use by employees upon retirement
**personal exemption** the value of merchandise that a person is allowed to bring into the country before any tax is added
**personal income tax return** a government form that is an official record of money earned during the year and the amount of income tax to be paid; it must be sent to the Canada Customs and Revenue Agency along with any tax due.
**Personal Tax Credits Return** a TD1 form, or a government form completed by new employees to assist employers in determining deductions from pay
**piecework** a job where payment is based on the number of units made, sold, or picked
**points program** an incentive plan where points are accumulated with each purchase and later redeemed for merchandise or discounts off of future purchases
**property tax** a municipal government tax on real estate (land with or without buildings) based on the value of the real estate
**PST (Provincial Sales Tax)** a provincial government tax on most goods and many services, 8% of the price in Ontario in 2002
**purchasing power** financial ability to make purchases

## R

**redeem (coupons or points)** use a coupon to obtain a discount or use points to obtain merchandise
**regular earnings** the total pay received on regular hours worked
**regular price** the original price that a retailer sets for an item
**rent to own** to pay a regular fee for a set time for an item which, if kept for the full time, the renter owns
**resume** a document that summarizes personal information, education and training, and work experience; it is submitted by an individual who is applying for a job.
**return (on an investment)** money made by selling for more than was paid and by receiving interest or dividends
**rewards (of a rewards program)** merchandise or discounts available by purchasing particular items or items in particular stores
**risk tolerance** the degree of uncertainty about the safety of an investment that an investor can tolerate
**room and board** a fee that includes lodging and meals
**RRSP (Registered Retirement Savings Plan)** a plan to which workers may contribute money for their retirement; contributions are tax deductible and earnings are tax exempt until withdrawn.

## S

**Safety Standards Certificate** a certificate required to register a used motor vehicle as roadworthy; it is obtained at a licensed Motor Vehicle Inspection Station for a fee and is valid for 36 days.
**salary** pay for work done, usually based on an annual amount, but paid at regular intervals, such as semi-monthly and monthly
**salary plus commission** a method of pay for salespersons that combines a set amount and an amount based on a percent of the value of the items sold or the number of items sold
**sale price** the discounted price that a retailer sets for an item
**sales tax** a provincial or federal government tax on the retail price of certain goods and services
**savings account** a bank account which pays interest on the money deposited
**seasonal** existing for only part of the year
**security (of a loan)** what can be taken if a loan is not repaid
**self-service banking** automated banking, or banking done at a bank machine, over the Internet, or by telephone
**selling price** the price a customer pays for an item, either a regular price or a sale price
**service charge** a fee for services performed
**shares** portions of a corporation
**shift** the working time of a group of employees
**shift work** employment usually associated with such places as grocery stores, factories, and hospitals, that need to be staffed for more than eight hours a day

**simple interest** interest calculated on the principal using the formula $I = Prt$, where $I$ is the interest in dollars, $P$ is the principal in dollars, $r$ is the interest rate as a percent, and $t$ is the time in years

**SIN (Social Insurance Number)** a government-issued number used to cross-reference all income and deductions

**small-cap stocks** shares in small, not well-established corporations

**sponsor (of a rewards program)** a store or business that participates in a rewards program

**spreadsheet** a series of rows and columns into which data can be entered and manipulated, usually software

**stages of life (for investing)** age groupings that help to determine a person's risk tolerance—early career, established, pre-retirement, retirement

**standard of living** how well one lives based on wealth, literacy, access to medical care, employability, and rights

**Statement of Interest Income** a T5 information slip, or an official one-year record of the amount of interest earned on investments

**Statement of Remuneration Paid** a T4 information slip, or an official one-year record of income earned and income taxes deducted; it also shows CPP, EI, and other deductions that affect income taxes.

**step commission** money earned for selling a product or service, based on a percent of the value of the items sold or the number of items sold where the percent changes

**stipend** an allowance paid as a thank you for work done, usually much less than earnings for the same work

**store money** money that is issued by a store for making purchases at that store

**straight commission** money earned for selling a product or service, based on a percent of the value of the items sold or the number of items sold where the percent changes

## T

**T1 General Forms** income tax forms to be completed by people who file income tax returns

**T4 slip** Statement of Remuneration Paid, or an official one-year record of income earned and income taxes deducted; it also shows CPP, EI, and other deductions that affect income taxes.

**T5 slip** Statement of Interest Income, or an official one-year record of the amount of interest earned on investments

**TD1 form** a Personal Tax Credits Return, or a government form completed by new employees to assist employers in determining deductions from pay

**telephone banking** self-service banking done over the telephone

**tender** offer money to pay for a purchase

**term (of a GIC)** the length of time that a GIC is in effect

**time and a half** an hourly rate equal to 1.5 times the regular hourly rate

**timesheet** a record of the hours worked by an employee who is paid an hourly rate

**tip** a small gift of money given for services, such as being served in a restaurant

**total purchase price of a new car** the total amount paid for a new car, including the base price of the car, all costs, fees, and taxes

**transaction** a debit or credit in a bank account or by credit card

**transaction fee** a charge for a transaction

**TSE (Toronto Stock Exchange)** an exchange located in Toronto where stocks and bonds are bought and sold by brokers

## U

**union dues** deductions to pay for membership in a union

**union member** a worker who pays dues to an organization that negotiates earnings, benefits, and working conditions with the employer

**unit price** the cost of an item expressed per unit, such as per item, per 100 g, per L

**utilities** water, heat, and electricity

## W

**withdrawal** money taken out of a bank account

**work experience** previous jobs or volunteering that resulted in gaining skills for future employment

# Answers

## Chapter 1: Working and Earning

### 1.1 Finding a Job pp. 2–3

**Develop**

1. **a)** Answers will vary. **b)** Answers will vary.
2. Answers will vary. For example, use a computer or a fax machine at a public place, such as a library, or at a business with these office services.
3. **child-care worker** **a)** caring **b)** ECE certificate **c)** none stated **d)** salary of $29 500/year **e)** no
   **sales help** **a)** friendly, conscientious worker **b)** must be fluent in English and possibly a second language **c)** none **d)** $54 per shift plus commission **e)** Answers will vary.
   **entry-level clerk** **a)** good communication and organizational skills **b)** none stated **c)** experience with Excel and Word **d)** salary of $24 000 with a quarterly bonus of $1200 **e)** Answers will vary.
   **server** **a)** friendly personality **b)** none stated **c)** none stated **d)** minimum wage plus tips **e)** Answers will vary.

**Practise**

**4–6.** Answers will vary.

### 1.2 Salary pp. 4–5

**Practise**

1. **a)** Weekly Earnings: $384.62; Monthly Earnings: $1666.67 **b)** Weekly Earnings: $355.77; Monthly Earnings: $1541.67 **c)** Weekly Earnings: $476.92; Monthly Earnings: $2066.67 **d)** Weekly Earnings: $557.69; Monthly Earnings: $2416.67 **e)** Weekly Earnings: $432.69; Monthly Earnings: $1875 **f)** Weekly Earnings: $615.38; Monthly Earnings: $2666.67
2. **a)** Annual Salary: $24 960; Weekly Earnings: $480 **b)** Annual Salary: $29 120; Monthly Earnings: $2426.67 **c)** Annual Salary: $19 680; Weekly Earnings: $378.46 **d)** Annual Salary: $19 500; Monthly Earnings: $1625 **e)** Annual Salary: $21 600; Weekly Earnings: $415.38 **f)** Annual Salary: $26 000; Monthly Earnings: $2166.67
3. Pauline

**Skills Check**

**a)** $11.45 **b)** $23.16 **c)** $42.44 **d)** $253.35 **e)** $152.82 **f)** $47.85

### 1.3 Piecework pp. 6–7

**Practise**

1. $45
2. **a)** $300 **b)** $750
3. Answers will vary. For example, because of the size of the tree and experience
4. Answers will vary. For example, workers might do a poor job if rushing.
5. **a)** $140 **b)** 6000
6. Answers will vary. For example, because of weather conditions
7. Answers will vary. For example, a worker earns more, and more work is accomplished.

### 1.4 Hourly Rate and Overtime Rate pp. 8–10

**Practise**

1. **a)** Answers will vary. For example, to compensate for the personal time that the worker is missing out on **b)** 1.5 times the regular hourly rate **c)** 2 times the regular hourly rate
2. $22 750
3. $380.55
4. **a)** cell E2: Regular Hourly Rate × Regular Hours = 8.5 × 44
   cell F2: Hours Worked − Regular Hours = 44 − 44
   cell G2: Regular Hourly Rate × 1.5 = 8.50 × 1.5
   cell H2: Overtime Hours × Overtime Rate = 0 × 12.75
   cell I2: Regular Earnings + Overtime Earnings = 374 + 0
   **b)** George: $375.00; Chris: $485.00; Hanya: $700.00; Joel: $489.25
5. **a)** $2785 **b)** 102 hours

**Skills Check**

**a)** 5 **b)** 7.5 **c)** 4.8 **d)** 2.24 **e)** 3.28 **f)** 1.4

## 1.5 Career Focus: Restaurant Server pp. 11–12

1. Answers will vary. For example, Sunil and Moira might strive to be clean, outgoing, punctual, and organized.
2. **a)** They should make sure that they are available between 5 p.m. and 6 p.m. **b)** Lack of availability would suggest that they are not interested or reliable.
3. **a)** \$42.60 **b)** \$213 **c)** \$101.10 **d)** \$505.50
4. Answers will vary. For example, perhaps Sunil and Moira appeared clean and neat, arrived on time, reflected an outgoing nature, and seemed organized.
5. **a)** Answers will vary. **b)** Answers will vary.
6. Pasta of the Day and Soup of the Day
7. Would you like something to drink?
8. Answers will vary. For example, Pasta of the Day with apple pie and a coffee, not including taxes or a tip
9. Answers will vary. For example, the server might approach the cook about the issue.

## 1.6 Commission pp. 13–15

### Practise

1. \$2350
2. \$502.80
3. **a)** cell G3: Sum of Daily Sales = 425 + 422 + 431 + 370 + 429
   cell H3: 25% of Total Sales = 0.25 × 2077
   **b)** John: \$535.50; Hans: \$540; Adam: \$519.50; Rita: \$512
4. \$735
5. **a)** cell E2: 4% of Price × Number Sold = 0.04 × 38.50 × 12
   **b)** 105.85
   **c)** cell E7: sum of commissions for all the magazines
   **d)** \$361.85
6. **a)** Answers will vary. For example, straight commission pushes employees to sell more, but offers no backup salary if conditions restrict their ability to sell; salary plus commission pushes employees to sell merchandise and provides them with a set salary as well. **b)** Answers will vary. For example, individuals with outgoing, helpful personalities

## 1.7 Step Commission pp. 16–17

### Develop

1. **a)** December, Christmas **b)** June. Answers will vary. For example, he might have been on vacation. **c)** 3
2. **a)** Shamir sold \$5175 over \$20 000; therefore, he receives 8% commission on that amount.
   **b)** Monthly salary is \$900. 5% commission on sales less than or equal to \$20 000 applies. 8% commission on sales more than \$20 000 is 8% of \$5175.
3. **a)** \$900. Yes, because \$900 is the set salary.
   **b)** January, February, and June because these months didn't have sales of \$20 000 **c)** January, February, and June **d)** January, February, and June because there were no sales over \$20 000
   **e)** December **f)** \$10 800. Yes, because it is 900 × 12. **g)** \$27 363.94
4. cell D6: 5% of \$20 000
   cell E6: sales − \$20 000 = 28 317 − 20 000
   cell F6: 8% of sales over \$20 000 = 0.08 × 8317
   cell G6: salary + 5% commission + 8% commission = 900 + 1000 + 665.36
   cell G14: sum of Total Earnings for January to December

### Practise

5. **a)** cell D2: 12% of sales = 0.12 × 4825
   cell E2: sales less than \$5000, so 0
   cell F2: 15% of sales over 5000 = 0.15 × 0
   cell G2: salary + 12% commission + 15% commission
   **b)** Jan.: \$1579.00; Feb.: \$1331.80; Mar.: \$1718.40; Apr.: \$1667.80; May: \$1697.60; June: \$1758.60; July: \$1700.80; Aug.: \$1641.20; Sept.: \$1742.40; Oct.: \$1736.60; Nov.: \$1658.00; Dec.: \$1836.60; Total: \$20 068.80

## 1.8 Putting It All Together: Finding a Job pp. 18–19

1–10. Answers will vary.

## 1.9 Chapter Review pp. 20–21

1. **a)** salary, piecework, hourly rate, commission **b)** hourly rate **c)** commission
2. **a)** \$515 **b)** \$2231.67
3. 50

4. **a)** cell E2: Regular Hourly Rate × Regular Hours = 7.50 × 44
cell F2: Hours Worked − Regular Hours = 45 − 44
cell G2: Regular Hourly Rate × 1.5 = 7.50 × 1.5
cell H2: Overtime Hours × Overtime Rate = 1 × 11.25
cell I2: Regular Earnings + Overtime Earnings = 333.00 + 11.25
**b)** Isobel: $432.25; Denise: $667.80; Peter: $425.00; Sara: $375.88
5. $925
6. **a)** $385 **b)** $420 **c)** Answers will vary. For example, Gerry earns more money and more work is accomplished.
7. $267
8. **a)** $20 **b)** $250 **c)** $187.50 **d)** $350
9. $2129.17
10. **a)** cell D2: 6% of 1000 = 0.60 × 1000
cell E2: Sales − $1000 = 1250 − 1000
cell F2: 10% of $250 = 0.10 × 250
cell G2: salary + 6% commission + 10% commission = 300 + 60 + 25
**b)** week 2: $349.00; week 3: $355.00; week 4: $370.00
11. $17.58/hour

# Chapter 2: Deductions and Expenses

## 2.1 Standard Deductions pp. 24–26

### Develop

1. **a)** Kevin J. Wilmont **b)** June **c)** 343 535 444
2. Answers will vary.
3. **a)** $262.50 **b)** $224.70

### Practise

4. **a)** Answers will vary. For example, health care, education, police, postal service, and defence
**b)** Answers will vary. For example, look it up in the blue pages of the phone book.
5. **a)** $23.50 **b)** $1222
6. $307.32
7. $436.28
8. **a)** income tax **b)** Answers will vary. For example, many government services are paid for by income taxes, while EI and CPP each have a specific purpose.

### Skills Check

**a)** Answers will vary. **b)** Answers will vary.
**c)** Answers will vary. **d)** Answers will vary.
**e)** 3.2% **f)** 2.25%

## 2.2 Other Deductions pp. 27–29

### Practise

1. EI, CPP, income tax, company pension plan, health plan, life insurance plan, disability insurance plan, union dues, payroll savings plan, and charitable donations
2. **a)** benefit **b)** benefit **c)** deduction
3. Total deductions: $93.03 Net income: $333.97
4. Total deductions: $79.57 Net pay: $292.43
5. **a)** $214.71 **b)** $385.07 **c)** $81.70 **d)** $498
6. **a)** More deductions are taken off a higher gross income. **b)** He should consider his net income because it is the amount of money that he actually receives.
7. **a)** $287.28 **b)** $76.66 **c)** $210.62
8. **a)** $402.50 **b)** $88.19 **c)** $314.31
9. $12.50/hour

## 2.3 Career Focus: Amusement Park Worker pp. 30–31

1. **a)** $360.75 **b)** Answers will vary. For example, for 22 weeks, $7936.50
2. **a)** $296.61 **b)** 82%
3. Less than, because his EI deductions are less than the amount he is paid per hour
4. Answers will vary. For example, mechanical, reading, writing, and decision-making skills
5. **a)** Pamela Grey **b)** 451 **c)** 75¢
6. Stop number − start number = number of tickets sold
7. **a)** Total sales: $338.25; Bills: $30.00; $50.00; $80.00; $50.00; $0.00; Coins: $0.05; $0.20; $0.50; $4.50; $49.00; $74.00; Total Bills: $210.00; Total Coins: $128.25; Total Cash: $338.25 **b)** yes
8. Answers will vary. For example, mechanical, mathematical, and people skills

## 2.4 Living Expenses p. 32

### Develop

1. **a)** Answers will vary. For example, groceries, rent, health care, phone, cable, transportation, clothes, and entertainment **b)** Answers will vary. **c)** Answers will vary.
2. Answers will vary.
3. Answers will vary.
4. Answers will vary.

## 2.5 Comparing Expenses pp. 33–34

### Develop

1. **a)** Answers will vary. **b)** Answers will vary.
2. **a)** Wearing uniforms that are paid for decreases expenses by not having to buy suitable clothes for work, but might increase expenses somewhat depending on whether the person otherwise wears clothes that require dry cleaning. **b)** decreases expenses **c)** decreases expenses
3. **a)** Cameron has fewer expenses to plan for than Sharon, because three are included in his rent. **b)** Felica's expenses are lower than Jackson's because she doesn't have to pay for her glasses. **c)** Joel's expenses are lower than Trevor's because Trevor needs transportation for 24 km each day he works. **d)** Gina's expenses could be higher because she has to dress better for work than her sister. **e)** Tricia's expenses are lower than Fatima's because Fatima has expenses related to her child.

### Practise

4. Answers will vary. For example, ability to buy things
5. Answers will vary.
6. **a)** Wealth, literacy, and employability are influenced by education, training, and work experience; with more education, you are eligible for more and better paying jobs and as part of an education, you learn to read. **b)** All measures give Canadians a higher standard of living. Canadians are wealthier, are more literate, are more employable, have better access to health care, and have more rights recognized compared to people in many developing nations.

### Skills Check

**a)** $7.50 **b)** $47.40 **c)** $48 **d)** $161 **e)** $217.50 **f)** $271.15

## 2.6 Paying Expenses pp. 35–37

### Develop

1. The amount paid in rent includes heat, electricity, and water.
2. **a)** $201 **b)** $352
3. **a)** October 4 **b)** October 19
4. **a)** Answers will vary. For example, Emelio might avoid paying bills after their due dates to keep in good standing. **b)** Answers will vary. For example, by not paying before the due dates, Emelio keeps his money as long as he can.
5. **a)** October 15 before receiving his pay **b)** October 15 after receiving his pay
6. **a)** $0 **b)** Answers will vary. For example, Emelio could wait till October 15 to buy groceries.

### Practise

7. **a)** $1992 **b)** $1910
8. yes
9. 96%
10. 46% (including groceries, which are not really due, and cable, which is the same day)
11. **a)** same timing **b)** Credit card payment and car payment due dates would be met only if she had money left from September 15. **c)** All bill payments would be met if she had money from September 30.
12. Answers will vary. For example, either pay bills early, or save money until the due dates.
13. Answers will vary.

## 2.7 Purchasing Power pp. 38–40

### Develop

1. **a)** $304.50 **b)** $266.80 **c)** $1156.13
2. **a)** $889 **b)** $267.13

### Practise

3. **a)** Answers will vary. For example, phone and cable expenses will likely double, because the low amount that Josephine is paying suggests that she shares these expenses with her friend. **b)** Answers will vary. For example, $1185 (based on new rent, doubling of phone and cable costs, and other expenses staying the same)

4. **a)** \$445 **b)** No, she has no transportation expense now.
5. Answers will vary. For example, \$473.87 (based on answer to question 3)
6. Answers will vary. For example, perhaps Josephine realized that it was unlikely that she could get a job that would pay at least the amount in question 5 or more than she is paid now.
7. Owning and operating this car would cost her about \$450.
8. **a)** cell E2: Rate of Commission × Typical Sales = 0.04 × 4100
cell F2: Salary + Commission = 1000 + 164.00
**b)** Amy's: \$952.00; Marci's: \$1380.00; Expressions: \$942.50
9. She spends \$28/month on medical expenses.
10. **a)** cell G2: 15% of Gross Monthly Income = 0.15 × 1164.00
cell H2: Gross Monthly Income − Deductions = 1164.00 − 174.60
**b)** Amy's: \$809.20; Marci's: \$1173; Expressions: \$801.12
11. None. Only Marci's has an insignificantly higher take-home pay.
12. **a)** \$1600 **b)** \$1518.40 **c)** \$1873.08
13. **a)** deduction: \$240; net income: \$1360
**b)** deduction: \$227.76; net income: \$1290.64
**c)** deduction: \$280.96; net income: \$1592.12
14. deduction: \$318.42; net income: \$1554.66
15. All. They all have higher take-home pays.
16. **a)** Answers will vary. **b)** Answers will vary. For example, what hours, days, evenings, and weekends would she work? Do they vary from week to week? Would there be opportunity for advancement?

### 2.8 Putting It All Together: Deductions and Living Expenses p. 41

**1–8.** Answers will vary.

### 2.9 Chapter Review pp. 42–43

1. A new employee completes a TD1 form to determine how much is to be deducted for standard government deductions.
2. Income tax is deducted to pay for services provided by the federal and provincial governments.
3. **a)** Net pay is often called take-home pay because it is the amount that you actually take home.
**b)** Answers will vary. For example, company pension plan, health plan, life insurance plan, and disability insurance plan **c)** benefit
4. **a)** \$330.83 **b)** \$342.20 **c)** \$342.23
5. Rent, for example, depends on where a person is living.
6. Answers will vary. Examples given:
essential: rent, groceries, health care, phone
non-essential: cable, Internet, entertainment
7. Answers will vary. For example, one person earns \$1000 each week, has no children and few expenses; another person earns \$1000 each week, has children and many expenses. The first person has greater purchasing power.
8. **a)** \$1768 **b)** \$1758 **c)** yes **d)** 99%
9. Answers will vary. For example, weekly, semimonthly, and monthly
10. Jeff has less than \$70 left after living expenses. If he had a car, he might not need a bus pass. However, car expenses would be a lot more than his remaining money and savings from not buying a bus pass.

## Chapter 3: Paying Taxes

### 3.1 Information for Filing Income Taxes pp. 46–49

**Develop**

1. **a)** Answers will vary. For example, to make the process easier for the tax filers **b)** Answers will vary. For example, to ensure that the forms are filled out correctly
2. Answers will vary. For example, provide employees with accurate information on their earnings, which they can use in filing tax returns
3. Answers will vary. For example, provide people with accurate information on their investments, which they can use in filing tax returns
4. **a)** from the businesses that you paid **b)** Answers will vary. For example, to reduce the amount of income tax paid
5. **a)** \$15 333.36 **b)** **i)** \$1702.96 **ii)** \$461.50 **iii)** \$368.00 **iv)** \$0 **v)** \$0 **c)** from the charities to which she donated

6. **a)** in the mail, at a postal outlet, or on-line **b)** from an employer **c)** from the bank/financial institution **d)** from the business or person paid, in this case, the child's caregiver
7. Jeremy has two jobs.
8. **a)** Answers will vary. **b)** Answers will vary.
9. $89.75
10. Answers will vary. For example, since a SIN is unique but names are not, the number confirms the person's identity.

**Skills Check**

**a)** $2.80 **b)** 70¢ **c)** $1.40 **d)** $2.10 **e)** $2.40 **f)** 80¢ **g)** $1.60 **h)** $2.40 **i)** $3.20

## 3.2 Help with Income Taxes p. 50

**Develop**

**1–4.** Answers will vary.

## 3.3 Provincial and Federal Sales Taxes pp. 51–53

**Practise**

Estimates will vary.

1. **a)** $4.34 **b)** $35 **c)** $17.50 **d)** 70¢ **e)** $1.40 **f)** 70¢
2. **a)** $6.40 **b)** $32 **c)** $38.40 **d)** $1.60 **e)** $9.60 **f)** $7.20
3. estimate of PST: $2.40; estimate of total cost: $32.40
4. **a)** estimate of GST: $14; estimate of total cost: $214; GST: $13.90; total cost: $212.40 **b)** estimate of GST: $4.90; estimate of total cost: $74.90; GST: $5.04; total cost: $77.04 **c)** estimate of GST: $9.80; estimate of total cost: $149.80; GST: $9.66; total cost: $147.66 **d)** estimate of GST: $4.20; estimate of total cost: $64.20; GST: $3.92; total cost: $59.92
5. **a) i)** $9.10 **ii)** $10.40 **iii)** $149.50 **iv)** $9.09 **v)** $10.39 **vi)** $149.37 **b) i)** $6.30 **ii)** $7.20 **iii)** $103.50 **iv)** $6.30 **v)** $7.20 **vi)** $103.50 **c) i)** $2.80 **ii)** $3.20 **iii)** $46 **iv)** $3.14 **v)** $3.58 **vi)** $51.51 **d) i)** $3.50 **ii)** $4 **iii)** $57.50 **iv)** $3.22 **v)** $3.68 **vi)** $52.85 **e) i)** $1.40 **ii)** $1.60 **iii)** $19 **iv)** $1.15 **v)** $1.31 **vi)** $18.85 **f) i)** $1.40 **ii)** $1.60 **iii)** $23 **iv)** $1.28 **v)** $1.46 **vi)** $21.03
6. 15%, 7% + 8% = 15%
7. $89.95. Answers will vary. For example, $103.44 ÷ 1.15 = $89.95

**Skills Check**

**a)** 21% **b)** 21% **c)** 18% **d)** 38% **e)** 23% **f)** 25%

## 3.4 Other Forms of Taxation pp. 54–55

**Develop**

1. Answers will vary. For example, you could ask someone who pays property taxes or look in the blue pages of the telephone book.
2. **a)** $7.85 **b)** $30.50 **c)** $25
3. **a)** $21.25 **b)** $19.45 **c)** 52%
4. **a)** $3.26 **b)** $2.85 **c)** $46.81 **d)** $27.36 **e)** 14¢
5. Answers will vary. For example, high taxes discourage people from buying them.
6. **a)** 24.7¢ **b)** $9.88 **c)** $7.41 **d)** $10.99
7. **a)** 15% **b)** 14%
8. **a)** 22% **b)** 20%
9. Answers will vary. For example, for 68.5¢/L **a)** 15% **b)** 21% **c)** 36%

**Skills Check**

**a)** $4.72 **b)** $1.45 **c)** $2.87 **d)** $5.28 **e)** $2.08 **f)** $2.66

## 3.5 Career Focus: Gas Bar Attendant pp. 56–57

1. Answers will vary. For example, being friendly, polite, and helpful
2. **a)** $37.35 **b)** $2.65
3. **a)** $4.39 **b)** $6.45
4. **a)** CA6 Regular Unleaded **b)** 152 **c)** 198
5. **a)** 165.7 L **b)** that it has uniform width **c)** 32 808.6 L **d)** $23 917.47 **e)** $3280.86 **f)** $4822.86

## 3.6 Putting It All Together: Paying Taxes p. 58

1. **a)** Answers will vary. For example, PST, GST, and income taxes **b)** Answers will vary. For example, jewellery, cigarette, and gasoline taxes

2. Answers will vary. For example, we pay for services that government provides.

3–5. Answers will vary.

## 3.7 Chapter Review p. 59

1. **a)** information slips (T4 and T5) and receipts for deductions or credits to be claimed **b)** Information slips (T4 and T5) are issued by employers and financial institutions respectively; receipts come from the sources of the expenses. **c)** Answers will vary. For example, to provide accurate information and proof for the government

2. **a)** Answers will vary. For example, classes at a library or a business that specializes in helping **b)** Answers will vary. For example, classes are free or very inexpensive, but you must be able to go when they are offered; hiring a business saves doing the work yourself, but costs money.

3. **a)** Answers will vary. For example, PST, GST, property tax, and jewellery tax **b)** Answers will vary. For example, health care, education, highway maintenance, and police

4. **a)** PST: Provincial Sales Tax; GST: Goods and Services Tax **b)** Answers will vary. For example, PST is collected by the provincial government, is 8% of the selling price, and is not collected on services, only goods. GST is collected by the federal government, is 7% of the selling price, and is collected on services as well as goods.

5. Estimates will vary. **a)** Estimate of PST: $10; Estimate of GST: $8.75; Estimate of Total Cost: $143.75; Actual PST: $10.04; Actual GST: $8.79; Actual Total Cost: $144.33 **b)** Estimate of PST: $4; Estimate of GST: $3.50; Estimate of Total Cost: $57.50; Actual PST: $4; Actual GST: $3.50; Actual Total Cost: $57.49 **c)** Estimate of PST: $3.20; Estimate of GST: $2.80; Estimate of Total Cost: $46; Actual PST: $3.12; Actual GST: $2.73; Actual Total Cost: $44.83 **d)** Estimate of PST: $15.20; Estimate of GST: $13.30; Estimate of Total Cost: $218.50; Actual PST: $15.18; Actual GST: $13.29; Actual Total Cost: $218.26 **e)** Estimate of PST: $18.40; Estimate of GST: $16.10; Estimate of Total Cost: $264.50; Actual PST: $18.39; Actual GST: $16.09; Actual Total Cost: $264.37

6. **a)** With direct taxes, the exact amount of tax collected is evident. Indirect taxes are included in the selling price. **b)** Answers will vary. For example, PST and GST are direct taxes; cigarette and jewellery taxes are indirect.

7. **a)** $41.10 **b)** 33%

8. 3

9. It is by $6.17.

# Chapter 4: Making Purchases

## 4.1 Making Change pp. 62–63

### Practise

1. **a)** $5.30 **b)** $2.02 **c)** $2 coin and 2 pennies **d)** $2 — 2.01 — 2.02 **e)** Answers will vary. For example, to lessen the number of coins to be received

2. **a)** $10.50 **b)** $2.13 **c)** $2 coin, one dime, and 3 pennies **d)** $2 — 2.10 — 2.11 — 2.12 — 2.13

3. **a)** $40 **b)** $13.05 **c)** one $10 bill, one $2 coin, one $1 coin, and one nickel **d)** $10 — 12 — 13 — 13.05

4. **a)** $30 **b)** $8.08 **c)** one $5 bill, one $2 coin, one $1 coin, one nickel, and 3 pennies **d)** $5 — 7 — 8 — 8.05 — 8.06 — 8.07 — 8.08

5. **a)** $15 **b)** $3.50 **c)** one $2 coin, one $1 coin, and 2 quarters **d)** $2 — 3 — 3.25 — 3.50

6. **a)** $15.12 **b)** $1.25 **c)** one $1 coin and one quarter **d)** $1 — 1.25

7. **a)** $8.50 **b)** $0.01 **c)** one penny **d)** 0.01

8. **a)** $10.14 **b)** $1 **c)** one $1 coin **d)** $1

9. **a)** $15.37 **b)** $3 **c)** one $2 coin and one $1 coin **d)** $2 — 3

10. **a)** $16 **b)** $0.39 **c)** one quarter, one dime, and 4 pennies **d)** $0.25 — 0.35 — 0.36 — 0.37 — 0.38 — 0.39

11. **a)** $20.45 **b)** $2.01 **c)** one $2 coin and one penny **d)** $2 — 2.01

12. **a)** one $1 coin, one quarter, one dime, and 2 pennies **b)** one quarter, one dime, and 2 pennies **c)** one dime and 2 pennies **d)** one nickel and 2 pennies **e)** 2 pennies

## 4.2 Getting Back Fewer Coins pp. 64–65

### Practise

1. **a)** 2 quarters, one nickel, and 3 pennies **b)** one $1 coin, 2 quarters, one nickel, and 3 pennies **c)** two

$2 coins, 2 quarters, one nickel, and 3 pennies **d)** one $10 bill, two $2 coins, 2 quarters, one nickel, and 3 pennies

**2.** **a)** 2 pennies **b)** one nickel and 2 pennies **c)** 2 dimes and 2 pennies **d)** 2 quarters, 2 dimes, and 2 pennies

**3.** **a)** $5.12 and $4.15 **b)** $5.51 and $5.01 **c)** $7.28 and $7.25

**4.** **a)** $6.40 **b)** $18.21 **c)** $24.35 **d)** $22.60

**5.** Answers will vary. For example, he might use the $10 bill if he had no interest in getting fewer coins or if he wanted coins. He might use the $10 bill, 2 quarters, and 3 pennies to get fewer coins back.

### Skills Check

**a)** $1.20 **b)** $4 **c)** 70¢ **d)** 85¢ **e)** 60¢ **f)** $2 **g)** 35¢ **h)** 42.5¢

## 4.3 Taxes and Total Cost pp. 66–68

### Practise

Estimates will vary.

**1.** **a)** $75 **b)** $2.25 **c)** $6 **d)** $10.50

**2.** **a)** Estimate of Total Taxes: $15; Estimate of Total Cost: $115; Actual Total Taxes: $14.93; Actual Total Cost: $114.43; Change Received: $5.57 **b)** Estimate of Total Taxes: $6; Estimate of Total Cost: $46; Actual Total Taxes: $6; Actual Total Cost: $45.95; Change Received: $4.10 **c)** Estimate of Total Taxes: $3; Estimate of Total Cost: $23; Actual Total Taxes: $3; Actual Total Cost: $22.98; Change Received: $2.02 **d)** Estimate of Total Taxes: $2.40; Estimate of Total Cost: $18.40; Actual Total Taxes: $2.44; Actual Total Cost: $18.73; Change Received: $1.30

**3.** **a)** $137.28 **b)** $12.72

**4.** **a)** Total cost before taxes: $12; Total taxes: $1.80; Total cost after taxes: $13.80 **b)** Total cost before taxes: $11.58; Total taxes: $1.74; Total cost after taxes: $13.32 **c)** $3.32

**5.** **a)** Total cost: $14.71; Change: $5.29 **b)** Answers will vary. For example, the price is 100% and the two taxes are 15%, giving 115%. **c)** 108%, Price (100%) + PST (8%) **d)** 107%, Price (100%) + GST (7%)

**6.** **a)** $3 **b)** Answers will vary.

**7.** Yes, the after-tax cost is only $9.76.

**8.** **a)** $4.27 **b)** $15.34 **c)** 63¢

### Skills Check

**a)** $4 **b)** $260 **c)** $9.60 **d)** $6 **e)** $210 **f)** $2.10

## 4.4 Discounts and Sale Prices pp. 69–72

### Develop

**1.** Answers will vary.

**2.** **a)** 80% **b)** 70% **c)** 85%

**3.** **a)** $70 **b)** $21

**4.** **a)** $70 **b)** $29.54

**5.** **a)** $33.33 **b)** $39.99 **c)** $46.66 **d)** $53.33 **e)** $59.99

### Practise

Estimates will vary.

**6.** $10.24

**7.** store with 10% off

**8.** **a)** Estimate of Sale Price: $210; Actual Sale Price: $211.49 **b)** Estimate of Sale Price: $550; Actual Sale Price: $551.65 **c)** Estimate of Sale Price: $290; Actual Sale Price: $287.98 **d)** Estimate of Sale Price: $400; Actual Sale Price: $399.20 **e)** Estimate of Sale Price: $48; Actual Sale Price: $46.75 **f)** Estimate of Sale Price: $20; Actual Sale Price: $21.20

**9.** **a)** $47.99 **b)** $123.66 **c)** $3599.10

**10.** coffee maker: $31.49; coffee grinder: $25.12

**11.** $64.99

**12.** **a)** $183.33 **b)** Answers will vary. For example, lack of after-sale service

**13.** Answers will vary. For example, sale price is $\$100 \times 0.75 \times 0.85 \times 0.8 = \$51$, so the discount is 49%.

## 4.5 Career Focus: Sales and Merchandising Clerk pp. 73–74

**1.** Answers will vary.

**2.** Answers will vary.

**3.** **a)** Method 1: $9.24 + (0.35 \times 9.24) = \$12.47$, so $12.49
Method 2: $1.35 \times 9.24 = \$12.47$, so $12.49
**b)** Answers will vary. For example, rounding up to the next 9¢ increases the price slightly, but seems lower than the next higher ten; $12.49 looks less than $12.50, but is only 1¢ less. **c)** sandpaper: $0.29;

measuring tape: \$3.19; door stop: \$1.09; window cleaner: \$2.09; picture wire: \$1.19; pack of sponges: \$0.79; air freshener: \$1.69; plant food: \$3.79; broom: \$2.09; package of tacks: \$0.49

4. **a)** one \$1 coin, 2 dimes, and 2 pennies **b)** one \$1 coin and 2 quarters **c)** one \$5 bill and 2 pennies **d)** one \$5 bill, one \$2 coin, one quarter, 2 dimes, and 3 pennies **e)** one \$1 coin and one dime

### 4.6 Sales Prices, Taxes, and Total Cost pp. 75–76

**Practise**

1. Estimates will vary. **a)** Estimate of Sale Price: \$5.40; Estimate of Total Cost: \$6.21; Actual Sale Price: \$5.30; Actual Total Cost: \$6.10 **b)** Estimate of Sale Price: \$13.60; Estimate of Total Cost: \$15.64; Actual Sale Price: \$13.42; Actual Total Cost: \$15.43 **c)** Estimate of Sale Price: \$32; Estimate of Total Cost: \$36.80; Actual Sale Price: \$34.36; Actual Total Cost: \$39.51 **d)** Estimate of Sale Price: \$34; Estimate of Total Cost: \$39.10; Actual Sale Price: \$33.33; Actual Total Cost: \$38.33 **e)** Estimate of Sale Price: \$51; Estimate of Total Cost: \$58.70; Actual Sale Price: \$50.98; Actual Total Cost: \$58.63
2. \$17.24
3. No. Answers will vary. For example, $(79.95 \times 0.75) + (8.50 \times 0.75) = 66.34$ and $66.34 \times 1.15 = 76.29$
4. The taxes are calculated on the sale price, not the original price, and the taxes, 15% of the sale price, are less than 15% off \$100.

### 4.7 Putting It All Together: Discounts, Taxes, and Total Cost p. 77

Answers will vary.

### 4.8 Chapter Review pp. 78–79

Estimates will vary.

1. **a)** \$10 **b)** \$3.19 **c)** one \$2 coin, one \$1 coin, one dime, one nickel, and 4 pennies **d)** \$2 — 3 — 3.10 — 3.15 — 3.16 — 3.17 — 3.18 — 3.19
2. **a)** \$20.07 **b)** \$2.75 **c)** one \$2 coin and 3 quarters **d)** \$2 — 2.25 — 2.50 — 2.75
3. **a)** \$15 **b)** \$1.21 **c)** one \$1 coin, 2 dimes, and one penny **d)** \$1 — 1.10 — 1.20 — 1.21
4. **a)** \$5.21 **b)** \$1.50 **c)** one \$1 coin and 2 quarters **d)** \$1 — 1.25 — 1.50
5. one \$5 bill, one quarter, and 2 pennies
6. one \$10 bill and one dime or one \$10 bill and two nickels
7. **a)** Estimate: \$57.50; Actual: \$55.19 **b)** Answers will vary.
8. **a)** Estimate of Total Taxes: \$7.50; Estimate of Total Cost: \$57.50; Actual Total Taxes: \$7.19; Actual Total Cost: \$55.14; Change Received: \$5.01 **b)** Estimate of Total Taxes: \$4.50; Estimate of Total Cost: \$34.50; Actual Total Taxes: \$4.49; Actual Total Cost: \$34.44; Change Received: \$5.56 **c)** Estimate of Total Taxes: \$0.45; Estimate of Total Cost: \$3.45; Actual Total Taxes: \$0.49; Actual Total Cost: \$3.78; Change Received: \$0.25 **d)** Estimate of Total Taxes: \$9; Estimate of Total Cost: \$69; Actual Total Taxes: \$8.83; Actual Total Cost: \$67.72; Change Received: \$2.28
9. \$21.74
10. **a)** Estimate of Sale Price: \$27; Actual Sale Price: \$31.49 **b)** Estimate of Sale Price: \$34; Actual Sale Price: \$38.24 **c)** Estimate of Sale Price: \$40; Actual Sale Price: \$43.97
11. discount store
12. **a)** \$39.38 **b)** 50%
13. **a)** Estimate of Sale Price: \$27; Estimate of Total Cost: \$31.10; Actual Sale Price: \$23.39; Actual Total Cost: \$26.90 **b)** Estimate of Sale Price: \$180; Estimate of Total Cost: \$207; Actual Sale Price: \$179.33; Actual Total Cost: \$206.23 **c)** Estimate of Sale Price: \$32; Estimate of Total Cost: \$36.80; Actual Sale Price: \$31.16; Actual Total Cost: \$35.83
14. The total cost is less than \$75 because the taxes, 15% of the sale price, are less than 15% off the regular price.
15. \$3.65

## Chapter 5: Buying Decisions

### 5.1 The Best Buy pp. 82–84

**Practise**

1. **a)** \$1.30/kg **b)** \$0.008/g **c)** \$0.82/100 g **d)** \$0.50/bag **e)** \$0.27/can
2. **a)** 4 L: \$0.87/L; 2 L: \$1.45/L; 1 L: \$1.79/L **b)** 4 L **c)** Answers will vary. For example, consider

whether or not you would use 4 L of milk before its Best Before date.

3. **a)** 1.950 kg package: \$5.241/kg; 0.620 kg package: \$5.484/kg **b)** Answers will vary. For example, the meat is priced differently by quantity. **c)** Answers will vary. For example, if you're not feeding many people, you might buy the smaller quantity with the higher unit price.
4. **a)** 1.6 L **b)** boxes: \$1.62/L; can: \$0.949 **c)** can **d)** Answers will vary. For example, consider where the apple juice is to be drunk.
5. \$92.95 ÷ 44.99 = 2.07, so 3 visits
6. 4

**Skills Check**

**a)** \$29.61 **b)** \$80.49 **c)** \$24.09 **d)** \$100.61 **e)** \$31 625 **f)** \$45.97

## 5.2 Incentives to Buy p. 85

**Develop**

1. Answers will vary. For example, newspapers, magazines, and flyers
2. **a)** 40¢ **b)** \$18.75 **c)** \$5 **d)** \$10 **e)** \$1.50
3. **a)** 1000 **b)** 6000 **c)** 750 **d)** 10 000
4. \$200
5. Answers will vary. For example, 180 000 points or 30 000 points and \$13
6. Answers will vary. For example, **a)** free **b)** 10 points **c)** cigarettes and tobacco, gifts with purchases, gift certificates, lottery tickets, products purchased at post office **d)** special signs in store **e)** \$5; 35% off; 13 000; 70% off; \$55; 34 000; \$75 **f)** When you have reached the required point level, you can get instant savings at the cash register.
7. **a)** Answers will vary. **b)** Answers will vary. **c)** Answers will vary. For example, entertainment tickets, merchandise, and travel
8. Answers will vary. Examples given: **a)** 22.5% **b)** on your credit card account
9. Answers will vary. Examples given: **a)** Money is tracked electronically. There is no minimum amount; just say you want to redeem it at the checkout. **b)** no
10. **a)** 3500 **b)** 5250
11. Answers will vary. Examples given: **a)** no **b)** when you do banking or use the credit card **c)** When you have a minimum number of points, you indicate at the checkout that you want to redeem points and your card will be swiped.

**Skills Check**

**a)** \$140 **b)** \$210 **c)** \$330 **d)** \$445 **e)** \$515 **f)** \$620 **g)** \$1090 **h)** \$2535

## 5.3 Cross-Border Shopping pp. 88–89

**Practise**

1. **a)** \$24.64 **b)** \$7.54 **c)** \$46.16 **d)** \$431.06
2. **a)** Canada **b)** Canada **c)** United States **d)** Canada
3. **a)** Estimates will vary. For example, \$187.50 **b)** \$193.13 **c)** none **d)** none
4. **a)** Estimates will vary. For example, \$90 **b)** \$91.16 **c)** \$91.16 **d)** \$13.67
5. **a)** Estimates will vary. For example, \$97.50 **b)** \$100.43 **c)** none **d)** none
6. **a)** Estimates will vary. For example, \$975 **b)** \$1002.71 **c)** \$252.71 **d)** \$37.91

## 5.4 Career Focus: Musician pp. 90–91

1. guitar #1: \$1985.33; guitar #2: \$1651.61; guitar #3: \$1002.71
2. Answers will vary. For example, the work gives her experience, helps her make contacts, and helps her decide if she really wants to work with children.
3. \$63.26
4. **a)** 10¢ **b)** 9¢ **c)** second package **d)** 3098
5. **a)** \$150 **b)** \$1200 **c)** \$2480
6. **a)** Answers will vary. For example, it might benefit from publicity. **b)** 200 **c)** \$12 750
7. **a)** 250 **b)** 12.5
8. Answers will vary. For example, CD sales might help fund the tour.

## 5.5 Deciding Which to Buy pp. 92–93

**Develop**

1. **a)** Model B **b)** Models A and C **c)** Model C **d)** Model B

2. **a)** Model B **b)** Model C **c)** Model A **d)** Answers will vary.

3. Answers will vary. For example, the Internet or a dealer

4. **a)** Answers will vary. **b)** Answers will vary.

### Practise

5. Answers will vary. Examples given: **a)** size, colour, material, price **b)** price, digital or not, zoom lens or not, size **c)** size, price, style, ice maker or not **d)** size, material, colour, price **e)** price, itinerary, ports of call, entertainment **f)** price, speed, services, equipment needed to use

6. Answers will vary.

### Skills Check

**a)** $72 **b)** $600 **c)** $2080 **d)** $34 **e)** $21.50 **f)** $74

## 5.6 Options to Pay—Layaway p. 94

### Develop

1. **a)** $182.85 **b)** $18.29 **c)** $164.56

2. **a)** $143.75 **b)** $14.38 **c)** $129.37

3. **a)** $322 **b)** $48.30 **c)** $273.70

### Practise

4. Answers will vary. For example, if you can't come up with the money, you lose your deposit, and you don't get the item when you initially want it.

5. Before: $98.90 After: $148.35

## 5.7 Renting with an Option to Buy pp. 95–96

### Develop

1. **a)** $350.75 **b)** $702 **c)** $351.25, a bit more than double the cash price including taxes

2. **a)** $78 **b)** Answers will vary. For example, because Jasmine needs the TV for only two months

3. $272.75, including taxes

4. **a)** $312 **b)** $156 **c)** $194.75, including taxes **d)** $506.75

### Practise

5. **a)** $1199.45 **b)** $320 **c)** $2400, more than double the cash price including taxes

6. $1039.45

7. **a)** $660.10 **b)** $440 **c)** $1320, exactly double the cash price including taxes

8. **a)** $385 **b)** $192.50 **c)** $467.60 **d)** $852.60

9. Answers will vary. For example, renting to own allows you to use something based on monthly payments, but it costs twice as much as paying cash.

10. Answers will vary. For example, the television, the refrigerator, and the washer and dryer were all double the cash price, including taxes.

11. within the first three months; 100% of rental payments apply to the purchase.

### Skills Check

**a)** $2400 **b)** $40.83 **c)** $1190 **d)** $642.75 **e)** $56.94 **f)** $422.20 **g)** $5306.25 **h)** $1448.17 **i)** $153.83

## 5.8 Buying on an Instalment Plan p. 97

### Develop

1. **a)** $194.85 **b)** $83.25

2. **a)** $77.85 **b)** $18.25

3. **a)** $320.85 **b)** $153.25

4. Omar. His item was the most expensive.

### Practise

5. **a)** $2976.20 **b)** $427.20 **c)** $3481.20 **d)** $505

6. **a)** $458.85 **b)** $98.85 **c)** $569.85 **d)** $111

7. **a)** $1263.85 **b)** $203.85 **c)** $1500.85 **d)** $237

## 5.9 No Interest or Payments for a Specified Time p. 98

### Develop

1. **a)** $199.80 **b)** $999 **c)** $1198.80 **d)** $100.20

2. **a)** $136.80 **b)** $579 **c)** $715.80 **d)** $64.22

3. **a)** $169.80 **b)** $799 **c)** $968.80 **d)** $107.10

4. **a)** Answers will vary. Examples given:
**Layaway** Advantage: No extra fees; Disadvantages: You don't get the purchase until you can pay for it in full; if you don't pay for it, you lose your deposit.
**Renting with an option to buy** Advantage: You get the use of the item for a monthly fee; Disadvantage: Very expensive
**Instalments** Advantage: Use of the item immediately; Disadvantage: There is a deferral charge or interest.

**No interest or payments for a specified time** Advantage: You get use of the item immediately; Disadvantages: If you don't have the money before the specified time, you will pay interest from the time of purchase; it might be more difficult to save money to cover the total amount than to make monthly payments. **b)** Answers will vary.

### 5.10 Putting It All Together: Deciding What to Buy and How to Pay p. 97

**1–6.** Answers will vary.

### 5.11 Chapter Review pp. 100–101

1. **a)** 340 g pkg: \$0.01/g; 600 g pkg: \$0.008/g **b)** 340 g pkg: \$1.11/100 g; 600 g pkg: \$0.88/100 g **c)** 600 g pkg **d)** Answers will vary. For example, the amount of cheese needed
2. **a)** \$27.50 **b)** \$23 **c)** \$6.50 **d)** 5: \$2.30; 10: \$2.10; 25: \$1.80 **e)** Answers will vary. For example, she needs only 10 passes.
3. Answers will vary. For example, store money, which you use towards future purchases, and points cards, which allow you to buy things for free or at a discount when you have accumulated enough points
4. **a)** \$38.61 **b)** \$162.83 **c)** \$81.72
5. **a)** \$330 **b)** \$338.36 **c)** \$50.75
6. Answers will vary. **a)** colour, size, price, waterproof or not **b)** size, components, price, quality **c)** quality, size, price, gas or propane
7. **a)** \$803.85 **b)** \$149.85 **c)** \$974.85 **d)** \$171
8. **a)** \$173.30 **b)** \$1002.30 **c)** \$103.50
9. **a)** Option 1: \$0; Option 2: \$1956.23; Option 3: \$345; Option 4: \$34.95; Option 5: \$44.95 **b)** Answers will vary.

## Chapter 6: Banking Transactions and Saving Money

### 6.1 Banking Transactions pp. 104–6

**Develop**

1. You can only withdraw money because you don't have an account at that bank.
2. Automatic Banking Machine
3. Answers will vary. For example, it's self-service because you do not need a teller.
4. You couldn't withdraw and deposit money because you are not at a bank or a bank machine.
5. **a)** withdrawing money from an ATM, depositing money at an ATM, paying bills at an ATM, transferring money at an ATM, checking account balance at an ATM, writing a cheque, paying bills by telephone banking, transferring money by telephone banking, checking account balance by telephone banking, paying bills on-line, transferring money on-line, checking account balance on-line, direct deposit, automatic withdrawal, using debit card **b)** Answers will vary.
6. Answers will vary.
7. **a)** \$0 **b)** \$2.75
8. \$12.90
9. \$6.65
10. \$11.65
11. Answers will vary.

### 6.2 Career Focus: Security Guard p. 107

1. **a)** to be insured against taking money **b)** Type G driver's licence and firearm licence
2. **a)** 250 **b)** \$312.50
3. Answers will vary.

### 6.3 Bank Statements and Passbooks pp. 108–9

**Develop**

1. **a)** May 5, May 19; automatically deposited **b)** May 31 **c)** \$874.74 **d)** withdrawal **e)** withdrawal **f)** Add deposits and subtract withdrawals. **g)** The money came out of her account at her bank.
2. **a)** September 23 **b)** withdrawal **c)** withdrawal **d)** \$5.40 **e)** \$495.86 **f)** \$495.86 **g)** at a teller, because you cannot deposit coins in a machine and there is no Access Point information
3. **a)** \$462.98 **b)** \$208.35 **c)** a withdrawal by cheque **d)** His pay was automatically deposited; he made a withdrawal. **e)** \$11.65 **f)** Answers will vary. For example, reduce the number of withdrawals.
4. Answers will vary.

## 6.4 Types of Savings pp. 110–11

**Develop**

1. Answers will vary.
2. **a)** long-term GIC because you don't have access for a longer period of time **b)** Answers will vary. For example, 1 to 5 years
3. Answers will vary. For example, no load (no fees), 1% or $45 (whichever is greater)
4. **a)** 2000, because the fund was at its highest **b)** 2001, because the fund was at its lowest
5. **a)** You have access to your money, but there are transaction fees. **b)** You earn more interest, but there is no access to money until term is up. **c)** More money could be made, but you risk losing money as well.
6. Answers will vary. Examples given: **a)** GIC **b)** savings account **c)** mutual funds

**Skills Check**

**a)** $4 **b)** $9.60 **c)** $5 **d)** $3.75 **e)** $5.40 **f)** $8.75

## 6.5 Simple Interest pp. 112–14

**Practise**

1. **a)** $1350 **b)** $400 **c)** $75
2. **a)** $675 **b)** $225 **c)** $5175
3. **a)** $490.67 **b)** $9690.67
4. **a)** $105 **b)** $1605
5. **a)** $225 **b)** $160 **c)** $108 **d)** $5793
6. **a)** Answers will vary. **b)** 5 years: $5200; 7 years: $5120
7. **a)** Answers will vary. **b)** 2 years: $2200; 1 year: $4200
8. **a)** Answers will vary. For example, because they use your money **b)** Answers will vary. For example, because you don't have access to it
9. Answers will vary. For example, fees: none like GICs; interest: comparable to GICs; term: 10 year, but cashable annually; amounts: only certain amounts while GICs can be any amount

**Skills Check**

**a)** 83.13 **b)** 1078 **c)** 9450 **d)** 3750 **e)** 111.38 **f)** 561 **g)** 131.25 **h)** 245 **i)** 36.56

## 6.6 From Simple Interest to Compound Interest pp. 115–17

**Develop**

1. **a)** $1000 **b)** $50 **c)** $2000
2. **a)** cell C5: Interest Rate × Principal = 0.05 × 1000 cell D5: Principal + Interest = 1000 + 50 **b)** D5
3. **a)** It increases. **b)** $2653.33 **c)** $1653.33 **d)** 15
4. **a)** The interest earned the first year in the spreadsheet is the same as simple interest; each year after, it is more. **b)** The interest in the spreadsheet is greater by $653.33. **c)** The amount in the spreadsheet is greater by $653.33. **d)** 15
5. **a)**

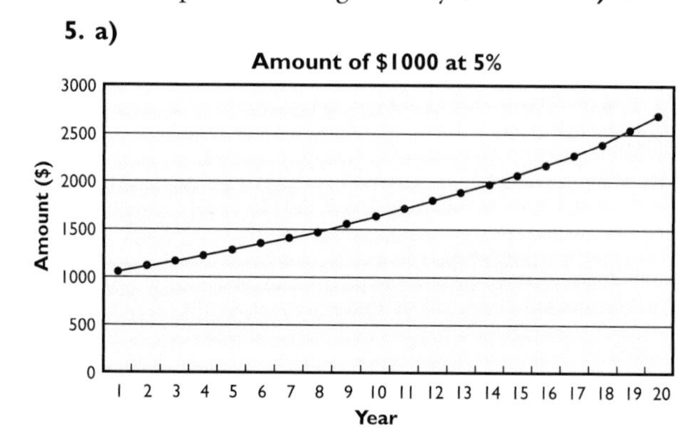


**b)** curve **c)** The amount increases as time increases, but it increases faster as more time passes.

6. **a)** $6000 **b)** $300 **c)** $11 000
7. **c)** It increases. **d)** $16 035.67 **e)** $11 035.67 **f)** 12
8. **a)** The interest earned the first year in the spreadsheet is the same as simple interest; each year after, it is more. **b)** The interest in the spreadsheet is greater by $5035.67. **c)** The amount in the spreadsheet is greater by $5035.67. **d)** 14
9. **a)**

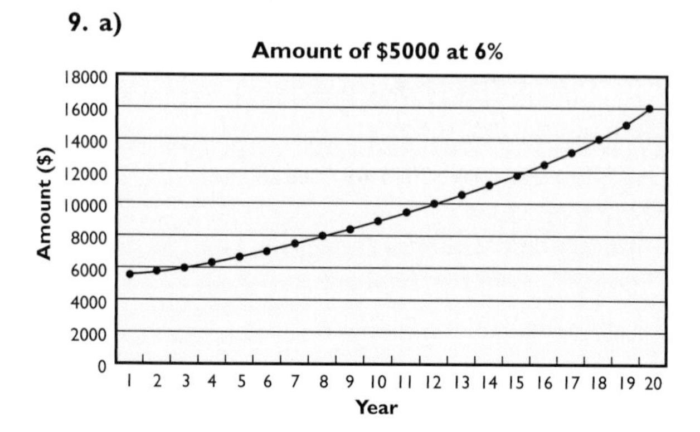

**b)** curves up to right, indicating the amount increases with time and increases faster as more time passes

**10. a)** $2800 **b)** $140 **c)** $4800

**11. c)** It increases. **d)** $7739.42 **e)** $5739.42 **f)** 11

**12. a)** The interest earned the first year in the spreadsheet is the same as simple interest; each year after, it is more. **b)** The interest in the spreadsheet is greater by $2939.42. **c)** The amount in the spreadsheet is greater by $2939.42. **d)** 13

**13.** Predictions will vary. For example, curves up to right

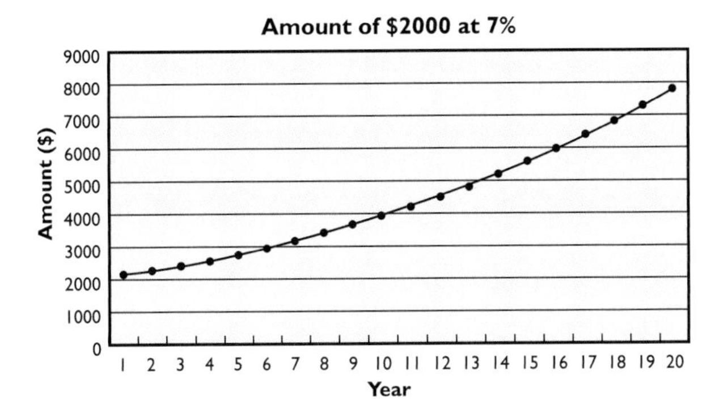


**14. a)** Answers will vary. For example, such interest increases over time. **b)** Answers will vary. For example, the amount from one year is used as the principal for the next year. **c)** Answers will vary. For example, with compound interest, you earn interest on interest. With simple interest, you earn interest on the original principal.

**Skills Check**

**a)** 116.25 **b)** 1.22 **c)** 1 **d)** 561.75 **e)** 5451.78 **f)** 16.01 **g)** 12 840 **h)** 82.74 **i)** 1327.83

## 6.7 Watching Savings Grow pp. 118–19

**Develop**

**1.** cell C5: Interest Rate × Principal = 0.06 × 100
cell D5: Principal + Interest = 100 + 6
cell B6: Amount at Next Birthday + Zarina Saved amount = 106 + 100

**2. a)** $4981.51 **b)** $2001.49 **c)** 28

**3.**

Investing $100 Each Year

Amount ($): 0, 1000, 2000, 3000, 4000, 5000, 6000

Number of Years: 1 2 3 4 5 6 7 8 9 10 11 12 13 14 15 16 17 18 19 20 21 22 23

**4. c)** $21 751.00 **d)** $14 769.50 **e)** $5013.28 **f)** 31

**5. c)** $11 435.36 **d)** $4310.11 **e)** $2195.61 **f)** 27

**6.** Answers will vary.

## 6.8 Compound Interest pp. 120–21

**Practise**

**1. a)** 1, 5, 6 **b)** 2, 3, 8 **c)** 2, 4, 4 **d)** 1, 7, 8 **e)** 4, 1.5, 12

**2. a)** $6700.48 **b)** $6333.85 **c)** $5849.29 **d)** $8590.93 **e)** $5978.09

**3. a)** $4875.98 **b)** $875.98

**4. a)** $13 685.69 **b)** $3685.69

## 6.9 Compounding Periods pp. 122–23

**Develop**

**1. a)** 1, 12, 2 **b)** 2, 6, 4 **c)** 4, 3, 8 **d)** 12, 1, 24 **e)** 365, $\frac{12}{365}$, 730

**2. a)** $1254.40 **b)** $1262.48 **c)** $1266.77 **d)** $1269.73 **e)** $1271.20

**3. a)** daily: $1271.20. Answers will vary. **b)** The principal increases more frequently, so the interest increases faster.

**Develop**

**4. a)** cell D2: Annual means 1 per year.
**b)** cell E2: Rate per Annum ÷ Compounding Periods in One Year = 12 ÷ 1
**c)** cell G2: Compounding Periods in One Year × Term (number of years) = 1 × 2
**d)** cell H2: $P(1 + i)^n = 1000(1.12)^2$

5.a)–g)

| | A | B | C | D | E | F | G | H |
|---|---|---|---|---|---|---|---|---|
| 1 | Principal, P ($) | Rate per Annum, % | Compounded | Compounding Periods in One Year | Rate per Compounding Period, i % | Term, years | Number of Compounding Periods, n | Amount, A ($) |
| 2 | 1000 | 12 | annually | 1 | 12 | 2 | 2 | 1254.40 |
| 3 | 1000 | 12 | semi-annually | 2 | 6 | 2 | 4 | 1263.48 |
| 4 | 1000 | 12 | quarterly | 4 | 3 | 2 | 8 | 1266.77 |
| 5 | 1000 | 12 | monthly | 12 | 1 | 2 | 24 | 1269.73 |
| 6 | 1000 | 12 | daily | 365 | 0.032876712 | 2 | 730 | 1271.20 |

6. **a)** daily: $1271.20. Answers will vary. **b)** The principal increases more frequently, so the interest increases faster.

## 6.10 Putting It All Together: Banking Transactions and Saving p. 124

**1–6.** Answers will vary.

## 6.11 Chapter Review p. 125

1. **a)** Answers will vary. For example, withdrawals, deposits, and transferring **b)** Answers will vary. For example, any two of depositing money, transferring money, paying bills, checking account balance **c)** A service charge is withdrawn from your account monthly.
2. **a)** May 21 **b)** May 9: withdrawal; May 28: withdrawal by a cheque he had written **c)** Add deposits and subtract withdrawals.
3. Answers will vary.
4. **a)** interest: $225; amount: $1725 **b)** interest: $400; amount: $4400
5. Compound interest accumulates faster because interest earns interest; interest from one period is added to the principal for the next.
6. **a)** 1, 5, 3 **b)** 2, 3, 4 **c)** 4, 2, 16
7. amount: $8337.11; interest: $1337.11

# Chapter 7: Investing Money

## 7.1 Types of Investments pp. 128–32

### Develop

1. **a)** They sell them for more than they bought, having earned interest. **b)** They lose money when the selling price and interest earned are less than the buying price.
2. **a)** They sell them for more than they bought, having earned dividends. **b)** They lose money when the selling price and dividends received are less than the buying price and commissions paid.
3. **a)** They sell them for more than they bought, having earned dividends or interest. **b)** They lose money when the selling price and interest earned and/or dividends received are less than the buying price and fees paid.
4. Answers will vary. For example, people without the time to monitor their investments or inexperienced investors
5. Answers will vary. For example, answers could be a cent different due to rounding.

### Practise

6. **a)** $700 **b)** $4400
7. **a)** $7655 **b)** She earned $17.
8. $1200
9. 414.938

**10. a)** $9039.28 **b)** $72.31 **c)** 1512.137

**11. a)** Bonds pay interest and are sold on stock market; market value is different from face value.
**b)** Stocks may pay dividends and are sold on stock market; they are most volatile. **c)** Mutual funds are a pool of bonds and shares, sold by brokers.

### Skills Check

**a)** $11 200 **b)** $4300 **c)** $17 400 **d)** $23 520
**e)** $4714.13 **f)** $2019.97 **g)** $2765.28 **h)** $3815.48

## 7.2 Registered Retirement Savings Plans pp. 133–35

### Develop

**1. a)** cell C4: Interest is compounded semi-annually.
**b)** cell C7: $P(1+i)^n = 2000(1.04)^2$
**c)** cell B8: Amount at Age 20 + Investment per Year = 2163.20 + 2000
**d)** There is no $2000 added; they are the same as C16, C17, and C18 respectively.

**2. a)** $20 000 **b)** 33 **c)** $531 818.79

**3. a)** $72 000 **b)** 53 **c)** $419 975.37

**4. a)** Jenna **b)** $111 843.42 **c)** Answers will vary.

### Practise

**5. a)** $341 214.72 **b)** $149 880.20 **c)** $90 542.79

**6.** Answers will vary. For example, considering Jenna and Leah, Jenna started early, invested $50 000 less, but will have over $100 000 more when she is 65.

## 7.3 Watching Investments pp. 136–37

### Develop

**1–3.** Answers will vary.

### Skills Check

**a)** $265 **b)** $7140 **c)** $13 180 **d)** $96 600 **e)** $840
**f)** $6100 **g)** $34 260 **h)** $1 284 000

## 7.4 Risk Tolerance pp. 138–40

### Develop

**1. a)** small-cap stocks **b)** corporate bonds

**2.** when you sell an investment for less than you paid for it

### Practise

Answers will vary. Examples given.

**3. a)** growth, safety, and income; she doesn't need income, and she can recover from losses because she has many years of work ahead. **b)** blue-chip stocks and corporate bonds for potential growth and not greatest risk **c)** Yes, she will need even more for a house so growth is important.

**4. a)** growth, safety, and income; she doesn't need income, and she can recover from losses because she has several years of work ahead. **b)** small-cap stocks and blue-chip stocks for potential growth

**5. a)** growth, safety, and income; he doesn't need income, and he can recover from losses because he has many years of work ahead. **b)** small-cap stocks, blue-chip stocks, corporate bonds, and government bonds; the variety provides growth and safety.

**6. a)** growth, safety, and income; she doesn't need to have income for a few years, and her ability to recover from losses is lower because she will work only 10 more years. **b)** corporate and government bonds and some blue-chip stocks for safety with some growth potential

## 7.5 Career Focus: Cook pp. 141–42

### Develop

**1.** eggs: 500, 2, 1000, 34
bacon: 375, 4, 1500, 75
sausages: 125, 3, 375, 3
whole wheat toast: 250, 2, 500, 25
white toast: 250, 2, 500, 25

**2. a)** Answers will vary. For example, these foods are likely to be found in other menu items as well.
**b)** Answers will vary. For example, coffee, cream, milk, sugar, butter, ketchup, salt, and pepper

**3. a)** $30 000 **b)** He needs growth; his goal is investment, not just savings.

**4.** growth, safety, and income; he wants to have money to buy the business in 15 years.

**5.** Answers will vary. Examples given: GICs: safe, but low interest, little growth; Canada Savings Bonds: safe, but low interest, little growth; government bonds: safe and some growth possible; corporate bonds: fairly safe and growth possible

blue-chip stocks: risky, but potential for growth
small-cap stocks: very risky, but greatest potential for growth

**6.** Answers will vary. For example, blue-chip stocks and corporate bonds

**7.** **a)** Answers will vary. For example, income, safety, and growth; since he will be retired, he cannot recover from losses and needs an income to supplement his Canada Pension. **b)** Answers will vary. For example, GICs and Canada Savings Bonds for safety and interest income, some corporate bonds for income, and a few blue-chip stocks for income and growth

## 7.6 Putting It All Together: Investing Money p. 143

**1–4.** Answers will vary.

## 7.7 Chapter Review pp. 144–45

**1.** **a)** $300 **b)** $4500 **c)** $1600

**2.** **a)** $2895 **b)** He earned $2278.

**3.** 152.284

**4.** Answers will vary. For example, advantages are the variety of investments and professional management; disadvantages are management fees and expenses and loss of control over investment decisions.

**5.** Trevor, because he started 20 years sooner than Reid

**6.** **a)** January–July 1999 **b)** $145 **c)** January–July 1997 **d)** $42

**7.** Answers will vary. For example, stock pages in newspaper and Web site

**8.** Income return is a dividend paid on interest earned, and a capital gain occurs when you sell for more than you paid.

**9.** **a)** the degree of fluctuation in your financial resources that you can withstand **b)** the amount of money available after expenses and the ability to recover from losses

**10.** Answers will vary. For example, **a)** income, safety of principal, and growth **b)** She should invest in a variety of bonds for income and safety.

**11.** **a)** established **b)** Answers will vary. For example, growth, safety, and income; he doesn't need income now and has potential to recover from losses. **c)** Answers will vary. For example, a mix of blue-chip stocks and bonds, with some low-cap stocks; he has potential to recover from losses.

**12.** **a)** $3656.98. Answers will vary. For example, no, or only part if rest is in something with more potential for growth **b)** $3825. Answers will vary. For example, no, or only part if rest is in something with more potential for growth **c)** $3900. Answers will vary. For example, yes, because she has the potential to recover from a loss and the bond pays interest **d)** Not possible to predict because stock could go up or down. Answers will vary. For example, yes, because she has the potential to recover from a loss and the stock pays a dividend **e)** Not possible to predict because stock could go up or down. Answers will vary. For example, no, due to higher risk and lack of dividend

# Chapter 8: Taking a Trip

## 8.1 Planning a Car Trip pp. 148–50

### Develop

**1.** Answers will vary.

**2.** **a)** Answers will vary. **b)** Weather could interfere with driving plans.

**3–11.** Answers will vary.

### Skills Check

**a)** 11:45 **b)** 2:45 **c)** 3:15 **d)** 11:45 **e)** 1:45 **f)** 10:45

## 8.2 Other Modes of Travel pp. 151–53

### Develop

**1.** **a)** $312.44 **b)** $321

**2.** **a)** to book it five days in advance **b)** yes

**3.** **a)** to book them together, seven days in advance **b)** $92.84

**4.** Answers will vary. For example, if a person is moving or already has transportation back, such as driving with someone

### Practise

**5.** **a)** $333.84 **b)** bus **c)** Answers will vary. For example, transportation to/from airport, train station, or bus terminal, food, and drinks **d)** Answers will vary. For example, comfort and duration **e)** Answers will vary. For example, by car, either own or rented

**6–8.** Answers will vary.

**9. a)** Answers will vary. For example, on business, to sightsee, to visit friends/relatives **b)** Answers will vary. For example, how soon they need to get to their destination, whether they want to make stops along the way

### Skills Check

**a)** 6:18 p.m. **b)** 9:27 a.m. **c)** 12:02 p.m. **d)** 12:14 a.m. **e)** 11:09 p.m. **f)** 9 hours, 33 minutes **g)** 2 hours, 55 minutes **h)** 3 hours

## 8.3 Reading Schedules pp. 154–57

### Develop

**1. a)** departure from Toronto at 6:40; departure from Moncton at 20:15 or 16:50 **b)** 19:00

**2. a)** 02/03/15, March 15, 2002 **b)** train 72 **c)** 1 hour, 41 minutes **d)** Answers will vary. For example, by 9:30

**3. a)** yes **b)** 9 hours, 25 minutes; 9 hours, 35 minutes; 8 hours, 50 minutes **c)** Answers will vary. For example, number of stops, speed travelled

**4. a)** 6203, 6205, 6209, or 6211 **b)** Answers will vary. For example, number of stops **c)** 07:00, 09:30, 11:30

### Practise

**5–9.** Answers will vary.

**10.** Answers will vary. Examples given: **a)** least expensive if you already have car, but takes longer **b)** fast, but costs more **c)** medium speed and medium costs, but more expensive than car and bus, and not as fast as plane **d)** less expensive, but slower than train and plane

## 8.4 Travelling Abroad pp. 158–59

### Practise

**1. a)** Danish krones **b)** Answers will vary.

**2. a)** Philippine pesos **b)** Answers will vary. **c)** Answers will vary.

**3.** Answers will vary. For example, **a)** 0.012 037; $30; $30.09 **b)** 1.4237; $50; $48.41 **c)** 0.204 517; $5; $5.11 **d)** 0.3576; $23; $25.03 **e)** 1.5950; $115; $119.63 **f)** 0.4030; $110; $100.75 **g)** 0.052 09; $60; $62.51

**4–5.** Answers will vary.

**6.** Less than 60 877.50 Japanese yen. Answers will vary. For example, $750 is the total cost, including taxes; shipping charges and duty do not apply, and transporting the component to Canada is easy and convenient to do.

**7.** Answers will vary. For example, cheques in Canadian and other currencies that you can buy, usually for a fee, and sign at time of purchase. Then you can treat them like money, signing them again when you use them. Traveller's cheques are safer than cash because if they are stolen you can have them cancelled. You pay the exchange rate on the day you purchase the cheques as with cash, while with a credit card or a debit card, you pay the exchange rate on the day of transaction. This may be an advantage or disadvantage depending on exchange rate fluctuations.

## 8.5 Career Focus: Flight Attendant p. 160

**1. a)** safety **b)** Answers will vary. For example, ongoing training **c)** Answers will vary. For example, friendly, people-oriented, helpful, and quick-thinking under pressure

**2.** Answers will vary. For example, 6 hours, 2 minutes

**3.** Answers will vary.

## 8.6 Putting It All Together: Planning Your Trip p. 161

Answers will vary.

## 8.7 Chapter Review pp. 162–63

**1.** Answers will vary. For example, **a)** Highway 11 to North Bay, continue Highway 11 to near Orillia, Highway 12 into Orillia **b)** about 555 km **c)** $36.20 **d)** about 7 hours; driving at about 80 km/h **e)** two; $25 **f)** Winter driving conditions could make driving slower; traffic in summer might be heavier.

**2.** Answers will vary. Examples given: **a)** You have a reliable car and time isn't an issue. **b)** Money isn't an issue and you need to get there quickly. **c)** Comfort, cost, and time are all equal issues.

**3. a)** $237.54 **b)** $908.43 **c)** $773.61

**4. a)** $237.54 **b)** bus **c)** Answers will vary. For example, transportation to/from airport, train station, or bus terminal, food, and drinks **d)** Answers will vary. For example, comfort and speed

5. **a)** Sunday **b)** Answers will vary. For example, the demand for this flight isn't great. **c)** depart: 3:00 p.m.; arrive: 7:51 p.m. **d)** depart: 8:40 p.m.; arrive: 11:55 p.m. **e)** 4 hours, 45 minutes
6. Answers will vary.
7. Answers will vary.

## Chapter 9: Borrowing Money

### 9.1 Credit Cards pp. 166–68

**Develop**

1. **a)** when the applicant has lived at the current address less than 3 years **b)** Louise has lived at her current address for 10 years, 4 months.
2. Answers will vary. Examples given: **a)** to be able to locate you **b)** to be able to check if you left where you lived for non-payment of rent or mortgage **c)** for identification purposes **d)** to determine your means of making money **e)** to know how much money you earn **f)** to know how much money you spend each month for rent or mortgage **g)** to check how much money you have in your accounts **h)** to show that you agree to the terms and conditions
3. layaway, instalment plans, no interest or payments for a specified time, and rent to own
4. Visa, MasterCard, and American Express

**Practise**

5. **a)** Type A: $0; Type B: $29; Type C: $99 **b)** Answers will vary. For example, the features of the cards, the minimum incomes, and the credit limits vary.
6. **a)** to know you have the potential to pay **b)** Students don't make as much money. **c)** all
7. Answers will vary. For example, they want to purchase something more expensive.
8. Answers will vary.
9. Answers will vary.
10. **a)** for security **b)** Answers will vary. For example, a couple
11. Answers will vary.

**Skills Check**

**a)** 2413.56 **b)** 11 572.00 **c)** 3612.22 **d)** 254.31

### 9.2 Delaying Payments on Credit Card Purchases pp. 169–71

**Develop**

1. **a)** $i$ is the rate per annum divided by the number of compounding periods per year. 0.186 is the rate per annum as a decimal, and 365 is the number of compounding periods in a year because it is compounded daily. **b)** $n$ is the number of compounding periods per year times the number of years. 365 is the number of compounding periods in a year and $\frac{1}{12}$ is the number of years because 1 month is $\frac{1}{12}$ of a year.
2. **a)** $132.19 **b)** $134.25 **c)** $2.06
3. **a)** the same **b)** Month 17; $9.22 **c)** 16 **d)** $19.22 **e)** yes

**Practise**

4. **d)** The minimum payment does not apply because the new balance to be paid is less. **e)** 25 **f)** $1.13 **g)** $51.13 **h)** yes
5. **d)** The minimum payment does not apply because the new balance to be paid is less. **e)** 44 **f)** $24.48 **g)** $274.48 **h)** yes
6. **d)** The minimum payment does not apply because the new balance to be paid is less. **e)** 49 **f)** $44.62 **g)** $830.62 **h)** yes

### 9.3 Short-Term Borrowing pp. 172–73

**Develop**

1. **a)** Answers will vary. For example, talk to someone at a bank. **b)** Answers will vary.
2. Answers will vary. For example, depending on how frequently they get paid
3. Answers will vary. For example, the greatest total amount you can borrow
4. **a)** Answers will vary. For example, talk to someone at a bank. **b)** Answers will vary. **c)** less **d)** Answers will vary. For example, a line of credit might be better because the interest rate is lower.
5. credit cards
6. **a)** With a line of credit, you are approved to borrow money before you even decide what you want to borrow it for. **b)** There is a credit limit, and that limit is the only buying restriction.

7. Answers will vary. For example, if they think they might occasionally have bills to pay before they receive their pay cheques

**Practise**

8. You don't have to put up security.
9. Answers will vary. Examples given: **a)** Delaying payments gives more time to pay them off, but costs more because of high interest rate to be paid. **b)** Loans allow you to make purchases when you need them and charge lower interest than credit cards, but cost more than paying cash. **c)** It is pre-approved for when you need it and the interest rate is lower than that of a credit card, but costs more than paying cash. **d)** The protection eliminates the worry of running out of money, but the high interest rate is like that of a credit card.
10. Interest rates are higher when you borrow. Banks make money paying less interest out than they charge.

**Skills Check**

**a)** 60 **b)** 14 **c)** 48 **d)** 72 **e)** 208 **f)** 156

## 9.4 Repaying Loans pp. 174–76

**Develop**

1. **a)** $2140 **b)** $8200
2. **a)** 12 payments per year; 1.9% annual interest rate; Amount of loan is $26 731.73. Length of loan is 3 years. **b)** $764.50 **c)** 12 payments per year for 3 years **d)** sum of the interest
3. **a)** 3 **b)** $27 521.97 **c)** monthly **d)** It decreases. **e)** It increases.
4. Predictions will vary.
5. **a)** Amount of each payment will increase; cost of the loan will increase. **b)** Amount of each payment will decrease; cost of the loan will decrease. **c)** Amount of each payment will decrease; cost of the loan will decrease. **d)** Amount of each payment will decrease; cost of the loan will increase.
6. **c)** $11 309.31
7. The more frequently the payments are made, the lower the amount of each payment and the lower the cost of the loan. Changing from monthly to semi-monthly payments has the greatest impact.
8. **a)** weekly **b)** amount of each payment: $188.97; cost of the loan: $9132.47

## 9.5 Career Focus: Small Business Ownership—Yard Maintenance pp. 177–78

1. Answers will vary. For example, wheelbarrow, twine, shears, clippers, rakes, leaf bags, lawn fertilizer, aerator, mulcher, trimmer, grass seed, cultivator, hammer, saw, screwdriver, pliers, gloves, and shovels
2. Answers will vary. For example, a line of credit might be a good choice because the interest rate is lower than for a credit card or overdraft protection, and they don't have to specify what they are going to buy. It gives them the flexibility to decide not to buy something like a cultivator if they don't have a big demand for it (they could just rent one as needed).
3. their expenses and what their competitors charge
4. Answers will vary. For example, check that they pay other bills on time and provide good service so customers will be satisfied
5. Answers will vary. For example, a copy of the bill with the date, name, address, and phone number of customer, work done, and price charged
6. Answers will vary. For example, so that they can contact them for the next year and know what they want done next
7. Answers will vary. For example, plough for their truck, snow shovels, and snow-blower

## 9.6 Putting It All Together: Borrowing Money p. 179

**1-5.** Answers will vary.

## 9.7 Chapter Review pp. 180-81

1. **a)** Answers will vary. For example, annual fee, minimum income required, credit limit, interest rate, and additional benefits **b)** annual fee and credit limit
2. **a)** loan, line of credit, and overdraft protection **b)** all **c)** overdraft protection **d)** line of credit **e)** loan **f)** loan and line of credit **g)** loan and line of credit; real estate and future wages
3. **d)** The minimum payment does not apply because the new balance to be paid is less. **e)** 38 **f)** $21.02 **g)** $221.02 **h)** yes

4. **c)** $11 831.19 **d)** Borrowing $10 000 and repaying over 5 years would have increased the loan cost. Having an interest rate of 7% and making payments weekly would have decreased the loan cost.

5. **a)** amount of each payment: $286.80; cost of the loan: $1766.33 **b)** Answers will vary. For example, he might see if he can get a lower rate of interest elsewhere, see if he can repay the loan over 5 years, see if he can buy the equipment he needs for less, or see if he can pay a lesser amount more frequently, for example, weekly.

# Chapter 10: Buying a Car

## 10.1 A Driver's Licence pp. 184–85

### Develop

1. You must maintain a zero blood alcohol level while driving, be accompanied by a fully licensed driver who has at least four years driving experience and a blood alcohol level of less than 0.05%, ensure that the accompanying driver is the only other person in the front seat, ensure that the number of people in the vehicle is limited to the number of working seat belts, refrain from driving on 400-series highways with a posted speed over 80 km/h or on high-speed expressways, and refrain from driving between midnight and 5:00 a.m.

2. You must maintain a zero blood alcohol level while driving and ensure that the number of people in the vehicle is limited to the number of working seat belts.

3. 2

### Practise

4. **a)** 400-series highways with a posted speed of over 80 km/h or high-speed expressways **b)** You need to learn how to drive on these highways.

5. **a)** You might need to retake a test if you don't pass one or if you remain a Class G1 driver for more than 5 years and need to renew. **b)** Answers will vary.

6. You can try the G1 Road Test four months sooner if you successfully completed a driver education course. Answers will vary. For example, insurance may cost less if you have successfully completed a course.

7. driver's licence or proof of passing knowledge test and a Class G vehicle to use

## 10.2 A New Car or a Used Car pp. 186–90

### Practise

1. Answers will vary. For example, **new** advantages: more choices, social appeal, warranty, likely lower maintenance costs, new safety features; disadvantages: high cost, high money payments, rapid depreciation in value
**used** advantages: lower cost, depreciation slower than for new, likely lower insurance costs; disadvantages: no or shorter warranty, fewer new car features

2. **a)** $30 712.45 **b)** $21 498.72

3. **a)** $23 045 **b)** $24 482.50

4. **a)** $21 703.50 **b)** Answers will vary. For example: Has it been in any accidents? Has it had any body work done? Has it been painted? Is there a list of service work done? What type of fuel consumption does it get? Can it be taken to a mechanic of his choice to get checked?

5. Answers will vary. For example, **from a dealer** advantages: three month warranty, additional warranty available for purchase; Safety Standards Certificate and the Drive Clean emissions test included; disadvantages: need to pay GST, licence fee, and fuel cost
**privately** advantages: doesn't have to pay GST, licence fee, and fuel cost; disadvantages: has to pay for the Safety Standards Certificate and the Drive Clean emissions test; no warranty

6. **a)** $13 875 **b)** $19 110 **c)** $11 330 **d)** $12 400

7. **a)** Answers will vary. For example, in 2002, $20 **b)** Answers will vary. For example, in 2002, $74 in Southern Ontario; $37 in Northern Ontario

8. Answers will vary. For examples, **new** costs before sales taxes: base price plus delivery, freight, pre-delivery inspection, pre-delivery expense, dealer preparation and/or administration fee; federal air conditioner tax, and fuel consumption tax; sales taxes: 8% PST and 7% GST; fees after taxes: possibly licence fee and fuel charge; advantages: more choices, social appeal, warranty, likely lower maintenance costs, new safety features; disadvantages: high cost, high money payments, rapid depreciation

**used from dealer** costs before sales taxes: price and maybe Safety Standards Certificate and Drive Clean emissions test; sales taxes: 8% PST and 7% GST; fees after taxes: possibly licence fee and fuel charge; advantages: three month warranty, additional warranty available for purchase, Safety Standards Certificate and the Drive Clean emissions test included; disadvantages: GST payment, licence fee, and fuel cost
**used privately** costs before sales taxes: price and maybe Safety Standards Certificate, Drive Clean emissions test, and some of costs involved to get the car to required standards; sales taxes: 8% PST on price of car and both on Safety Standards Certificate, Drive Clean emissions test, and some of costs involved to get the car to required standards, if applicable; advantages: no GST payment, licence fee, or fuel cost; disadvantages: no warranty, payment may be required for the Safety Standards Certificate, the Drive Clean emissions test, and some of costs involved to get the car to required standards

#### Skills Check

**a)** $16 396 **b)** $31 368 **c)** $25 053.60 **d)** $32 225
**e)** $19 139.20 **f)** $21 024.80 **g)** $23 720.80
**h)** $22 188.80

## 10.3 Buying Versus Leasing pp. 191–92

#### Develop

1. **a)** four **b)** $30 654.72 **c)** $4054.72
2. **a)** Answers will vary. For example, **loan** advantage: you own the car; disadvantage: higher monthly payments than a lease
   **lease** advantages: lower monthly payments than a loan, one way to get a new car every three or four years, and you don't have to trade it in; disadvantages: you pay for any damages and for any kilometres driven over the specified number in the agreement when the lease is up, and if you want to buy the car when the lease is up, you pay more than you would have if you had bought the car initially with a loan.
   **b)** Answers will vary.

#### Practise

3. **a)** $437 **b)** $519.80 **c)** $332.35 **d)** $462.99
4. **a)** $1990.35 **b)** $2619.70 **c)** $4331.28
5. **a)** five **b)** $23 980 **c)** $2460 **d)** $15 085.60
   **e)** $24 860.60 **f)** Answers will vary.

## 10.4 Career Focus: Car Salesperson pp. 193–94

1. $26 524.75
2. Answers will vary.
3. $405
4. **a)** cell D2: 30% of Dealer's Profit on Maureen's Deals = 0.3 × 9600
   cell E2: 6 cars sold per month means no bonus.
   cell F2: Commission Earned + Bonus Earned = 2880.00 + 0.00
   **b)** Jan.: $2880.00; Feb.: $2239.50; Mar.: $4630.40; Apr.: $4541.00; May: $8471.60; June: $9456.00; July: $2917.10; Aug.: $1902.00; Sept.: $4876.70; Oct.: $4957.10; Nov.: $6572.00; Dec.: $1047.00
   **c)** Jan., Feb., Aug., and Dec. because no bonuses were earned

## 10.5 Insuring a Car pp. 195–96

#### Develop

1. Answers will vary. For example, a $300 deductible has a higher premium because the insurance company would have to pay out more when a claim is made.
2. Answers will vary. For example, a new sports car
3. Answers will vary. Examples given: **a)** being 18, because you are a new driver and less experienced **b)** Toronto, because there is more likelihood of being in an accident where traffic is heavier
   **c)** last year, because you are inexperienced
   **d)** not taking it, because you may not be trained as well **e)** having four tickets, because it indicates a poor driving record **f)** suspended licence, because it indicates a bad driving record **g)** cancelled policy, because it indicates a bad driving record
   **h)** 30 000 km, because an accident is more likely to happen if you drive more

#### Practise

4. $1116
5. Answers will vary. For example, the insurance industry keeps statistics that help determine the likelihood of being in an accident and these show that sex, age, and marital status make a difference.
6. Answers will vary.
7. Answers will vary. For example, insurance is a large cost and you need to be able to afford it to drive your car.

## 10.6 Owning and Operating Costs pp. 197–98

**Practise**

1. **a)** and **b)** Answers will vary. For example, in 2002, \$74 in Southern Ontario and \$37 in Northern Ontario
2. \$43.38
3. \$460
4. \$943.17 monthly, 53.7¢/km
5. \$8605.18 annually, \$717.10 monthly
6. **a)** cell J2: Gasoline + Oil + Maintenance/Repairs + Insurance + Licence Sticker + Depreciation + (12 Monthly Car Payments) = 1800 + 114 + 1024 + 1250 + 74 + 1400 + (12 × 268)
   cell K2: Annual Cost ÷ 12 = 8878 ÷ 12
   cell L2: Annual Cost ÷ Distance Driven = 8878 ÷ 21 380
   **b)** Clare's: \$5490 annually, \$457.50 monthly, 33¢/km; Toma's: \$9100 annually, \$758.33 monthly, 61¢/km; Gilda's: \$10 745 annually, \$895.42 monthly, 40¢/km; Cale's: \$9698 annually, \$808.17 monthly, 53¢/km; Moufeed's: \$5822 annually, \$485.17 monthly, 31¢/km; Janet's: \$10 056 annually, \$838.00 monthly, 57¢/km
7. Answers will vary. For example, maintenance varies depending on the age and make of the car, insurance varies depending on the car and the driver, licence sticker varies depending on where you live, depreciation varies depending on the age and make of the car, and payments vary depending on total cost, interest rates, term of loan, and amount you can afford to pay.

## 10.7 The Costs of Irresponsible Driving pp. 199–200

**Develop**

1. **a)** four points **b)** failing to remain at the scene of a collision **c)** two years from date of offence **d)** nine points **e)** 15 or more points

**Practise**

2. \$98.75, three points
3. **a)** \$100 **b)** \$265
4. Answers will vary. For example, legal costs
5. Answers will vary. For example, insurance premiums will increase.
6. \$200
7. one

## 10.8 A Car Versus Public Transportation pp. 201–2

**Develop**

1. \$8000 annually, \$666.67 monthly
2. \$5600 annually, \$466.67 monthly
3. \$7400 annually, \$616.67 monthly
4. \$1446
5. **a)** 1, 3, 2, 4 **b)** Answers will vary. For example, schedule and convenience of bus relative to his needs **c)** Answers will vary. **d)** Answers will vary. For example, advantages include the ability to go where and when he wants to, and disadvantages may include the many costs and responsibilities related to running a car. **e)** Answers will vary. For example, advantages include less responsibility and less expense, and disadvantages include less freedom to travel when and where he wants.

**Practise**

6. **a)** GO/TTC Twin Pass **b)** \$62.50 **c)** single fare **d)** Answers will vary. For example, advantages include not having to drive in heavy traffic, sometimes being able to do such things as reading as you ride, and having fewer expenses because you don't have to pay for parking and operating a car. Disadvantages include having to travel where and when transportation is available.
7. Answers will vary. For example, Shair is concerned about the cost while Jeremy may want the convenience of having a car regardless of cost.
8. Answers will vary.

## 10.9 Putting It All Together: Buying a Car p. 203

Answers will vary.

## 10.10 Chapter Review pp. 204–5

**1. a)** 20 months **b)** G1 and G2 **c)** two
**d)** Answers will vary. For example, any three of the following: You must maintain a zero blood alcohol level while driving, be accompanied by a fully licensed driver who has at least four years driving experience and a blood alcohol level of less than 0.05%, ensure that the accompanying driver is the only other person in the front seat, ensure that the number of people in the vehicle is limited to the number of working seat belts, refrain from driving on 400-series highways with a posted speed over 80 km/h or on high-speed expressways, and refrain from driving between midnight and 5:00 a.m.
**e)** every five years, in 2002, $50

**2. a)** $20 713.95 **b)** $12 600

**3.** $19 813.50

**4.** $3643.85

**5.** Answers will vary. For example: Amount of deductible: The greater the deductible, the lower the premium. Value of car: The greater the value, the greater the premium. Driving record: The better the record, the lower the premium. Driving experience: The greater the experience, the lower the premium. Distance driven: The greater the distance, the greater the premium. Where you live and will be operating the car: The more traffic you drive in, the higher the premium.

**6.** $577.50

**7. a)** $13 282 **b)** $1106.83 **c)** 72.1¢

**8.** Answers will vary. For example, you pay fines and maybe legal fees, and your insurance premiums increase; you may injure or kill yourself or someone else, you could go to jail, and you might lose your job.

**9.** Answers will vary.

**10.** Answers will vary.

# Index

## Chapter 4

(page 77, bottom) Richard Hutchings/Corbis

## Chapter 5

(page 92) Tom King/The Image Bank/Getty Images

## Chapter 7

(page 133, left) Eyewire

(page 133, middle) Comstock Images

(page 133, right) Dick Luna/FPG International/ Getty Images

(page 137) Dick Hemingway

(page 138) Eyewire

## Chapter 8

(page 149) Jeff McIntosh/CP Photo Archive

(page 154) Terry Vine/Stone/Getty Images

(page 157) Rommel/Masterfile

(page 158) Susan Van Etten/PhotoEdit

(page 160) MaXx Images/Jeff Greenberg

(page 163) MaXx Images/Canstock Images Inc.

## Chapter 9

(page 179) Donald Nausbaum/Stone/Getty Images